Rehabilitation of Water Mains

AWWA MANUAL M28

Second Edition

American Water Works Association

MANUAL OF WATER SUPPLY PRACTICES—M28, Second Edition
Rehabilitation of Water Mains

Copyright © 1987, 2001 American Water Works Association

All rights reserved. No part of this publication may be reproduced or transmitted in any form or by any means, electronic or mechanical, including photocopy, recording, or any information or retrieval system, except in the form of brief excerpts or quotations for review purposes, without the written permission of the publisher.

Project Manager: Melissa Christensen
Technical Editor: David Talley
Production Editor: Carol Magin

Library of Congress Cataloging-in-Publication Data

Rehabilitation of water mains.-- 2nd ed.
   p. cm. -- (AWWA manual ; M28)
  Rev. ed. of: Cleaning and lining water mains. 1st ed. c1987.
  Includes bibliographical references.
  ISBN 1-58321-026-1
  1. Water-pipes--Maintenance and repair. 2. Water-pipes--Cleaning. 3. Water-pipes--Linings. I. AWWA Research Foundation. II. Title: Cleaning and lining water mains. III. Series.

TD491.A49 R44 2001
628.1'5'0288--dc21

   2001022658

Printed in the United States of America

American Water Works Association
6666 West Quincy Avenue
Denver, CO 80235

ISBN 1-58321-026-1

Printed on recycled paper

Contents

List of Figures, v

List of Tables, vii

Foreword, ix

Chapter 1 Distribution System Performance Criteria **1**

Water Quality, 1
Hydraulic Performance, 2
Structural Performance and Leakage, 3
Rehabilitation Solutions, 4
Selection of Rehabilitation Solutions, 4

Chapter 2 Cleaning **9**

Flushing, 9
Cable-Attached Devices, 10
Fluid-Propelled Cleaning Devices, 11
Cleaning by Power Boring, 17

Chapter 3 Lining Techniques **21**

Cement–Mortar Lining, 21
Epoxy Lining, 25
Slip-Lining, 28
Modified Slip-Lining Techniques, 32
Cured-in-Place Lining Techniques, 36

Chapter 4 Internal Joint Seals **41**

Fitting Procedure for Internal Joint Seals Installed in Water Mains, 41

Chapter 5 Pipe Bursting **47**

Chapter 6 Maintaining Service **51**

Bypass Piping, 51
Summary, 54

Chapter 7 Customer Relations **55**

Customer Notification Program, 55
Community Relations, 58
Summary, 58

Appendix A Structural Lining Design Issues **59**

Index, 63

List of AWWA Manuals, 65

This page intentionally blank.

Figures

1-1   Pipe with problems caused by corrosion, 2

1-2   Selection of rehabilitation techniques to resolve structural problems, 5

1-3   Selection of rehabilitation techniques to resolve water quality problems, 6

1-4   Selection of rehabilitation techniques to resolve flow, pressure, and leakage problems, 7

2-1   Drag cleaning, in which a winch pulls a mechanical cleaner through the pipe, 10

2-2   Foam pig: A bullet-shaped device made of polyurethane foam, 11

2-3   A foam pig with hardened coatings, 11

2-4   Loose debris flushed ahead of the pig, 12

2-5   Pigs launched through a disassembled fire hydrant for a 6-in. (150-mm) or smaller line, 13

2-6   Scraper unit with specially tempered steel blades, 14

2-7   Several scraper units assembled together in the field, 14

2-8   A series of disks to act as a hydraulic piston, pulling scrapers through the line, 15

2-9   A sandbag dam to create a pond for particle settling, 16

2-10   A spool piece installed at the entry and exit points for mechanical scrapers, 16

2-11   Rack-feed boring machine, 18

2-12   Cleaning pipe by power boring, 18

2-13   Cleaning head, 19

3-1   A cement–mortar lining machine for use in small-diameter pipe, 22

3-2   Introduction of a small lining machine, 23

3-3   A cement–mortar lining machine for use in large-diameter pipe, 23

3-4   A pipe ready to be returned to service four to seven days after cement–mortar lining, 25

3-5   A typical epoxy lining machine, 26

3-6   A typical epoxy lining application head, 27

3-7   An epoxy-lined water main, 27

3-8   HDPE pipe being formed into a C-shape on site, 33

3-9   Folded and banded HDPE pipe is winched into the host pipe, 35

3-10   Felt tube fed into the pipe, 37

3-11   Filling the tube-feeding standpipe with cold water, 38

3-12   Heating water after the tube is in place, causing the resin to adhere to the pipe walls, 38

4-1	Joint area is cleaned and prepared prior to installation of the seal, 43
4-2	A nontoxic lubricating soap is applied and the seal is carefully positioned with its retaining band, 43
4-3	An expansion ring is placed over each retaining band and a wedge is inserted between the band ends, 46
4-4	Each seal is leak tested twice before the main is put back into service, 46
5-1	Diagram of a typical pneumatic pipe-bursting operation, 48
6-1	Bypass installation for residential and commercial water service, 52
7-1	A sample letter notifying consumers of work to be done, 56
7-2	A sample caution notice to be posted at the work site, 57

Tables

1-1   Hazen-Williams coefficient, 3

4-1   Material details for internal joint seals, 42

4-2   Retainer band expansion pressures, 44

This page intentionally blank.

Foreword

This manual was written for water system designers, engineers, operators, and managers who need a reference guide for analyzing various water main rehabilitation options. This manual summarizes current information about water main rehabilitation technologies used in the United States and Europe today. All specified technologies must have NSF61 (National Sanitary Foundation) approval for components that come into contact with potable water.

While there are many emerging technologies that show promise for water main rehabilitation, this manual specifically addresses those technologies that 1) can be employed by a water utility to successfully rehabilitate water mains and 2) have a proven track record within the water industry. No attempt was made to evaluate one method of water main rehabilitation over another, and no attempt was made to evaluate relative costs between competing systems. It remains for the utility to decide which system will best suit a specific project, taking into account a variety of factors, including the cost of the process, the social disruption, the need to increase or maintain existing water main diameter, and so forth.

According to a US Environmental Protection Agency needs assessment survey, approximately \$80 billion may be required to improve transmission and distribution water mains to meet clean water requirements over the next decade. In the past this was accomplished by installing new facilities to replace aging infrastructure. With the exception of cleaning and cement–mortar-lining (which has been in use since the 1930s), the only option available to water utilities was costly dig-and-replace techniques. This manual describes viable, less costly options that have the potential to be less disruptive to the public, while providing the utility with a long-term solution to their water main needs.

The first edition of American Water Works Association (AWWA) Manual M28 was written by the AWWA Cleaning and Lining Committee. The second edition was written by the same committee; however, because of changes in the water main rehabilitation industry, the name of the committee was changed to the AWWA Water Main Rehabilitation Committee.

The Chairman of the committee, Michael E. Grahek, would like to thank the contributing authors responsible for new material supplied to expand the current manual into an up-to-date water main rehabilitation manual, specifically, Dr. John Heavens, Bernie Monette, Gary Zinn, Darrin Thomas, Benedict Ebner, Michael Landes, Daniel Moore, John Dugan, and George Mallakis.

In addition to those who contributed directly, the balance of the committee participated by reviewing and approving materials submitted for inclusion in the manual, adhering to the highest ethical standards, and keeping the needs of the water-using public at the forefront. The committee had the following members at the time:

Michael E. Grahek P.E., *Chairman*

Dave Arthurs, ARB Inc., Lake Forrest, Calif.
Tim Ball, Louisville Water Company, Louisville, Ky.
W.M. Breichner, Mainlining Service Inc., Hagerstown, Md.
Michael Cronin, Lineal Industries, Columbus, Ohio
J.M. Dugan, J. Fletcher Creamer & Son Inc., Hackensack, N.J.
B.H. Ebner, Heitkamp Inc., Watertown, Conn.
W.D. Ensor, Gannett Fleming Inc., Newport News, Va.
M.E. Grahek, Los Angeles Department of Water & Power, Los Angeles, Calif.
David Hughes, Philadelphia Suburban Water Company, Bryn Mawr, Pa.
Dr. John Heavens, PhD, Insituform Technologies Inc., Chesterfield, Mo.
Steven Kramer, Sverdrup Civil Inc., Falls Church, Va.
Mike Landes, Flowmore Services Corporation, Houston, Tex.
Bernie Monette, Doty Bros. Equipment Company, Norwalk, Calif.
Dan Moore, City of Greeley, Greeley, Colo.
Ken Morgan, P.E., Denver Water, Denver, Colo.
F.J. Moritz, Jr., Ridgewood Water Company, Ridgewood, N.J.
Stewart Nance, Raven Lining Systems, Tulsa, Okla.
E.O. Norris III, P.E., South Central Connecticut Regional Water
    Authority, New Haven, Conn.
C.K. Orton, TT Technologies Inc., Redwood City, Calif.
D.J. Smith, Water District No. 1 of Johnson County, Merriam, Kans.
J.P. Sullivan, Jr., Boston Water & Sewer Commission, Boston, Mass.
Darrin Thomas, Earth-Tech, Greensboro, N.C.
G.M. Zinn, Adtec-Merco, Merriam, Kans.

AWWA MANUAL M28

Chapter 1

Distribution System Performance Criteria

WATER QUALITY

The quality of treated drinking water may vary considerably, both from system to system and within a system, as a result of deterioration after it leaves the treatment plant and comes in contact with the interior of distribution system piping. Over time, changes in water chemistry can cause problems throughout the distribution system, ultimately affecting the quality of the water delivered to the end user.

The specific nature of distribution system water quality problems vary with water chemistry. However, the majority of these problems fall into three categories: sedimentation, encrustation, and fouling.

Sedimentation

Sedimentation is the process whereby solids settle out of water moving at low velocity in a main, reducing interior cross section and capacity. Source water pipelines or pipelines carrying improperly treated water can be subject to deposits of sand, silt, or organic materials. In extreme cases, sedimentation can also contribute to hydraulic problems, particularly at low points in the pipe.

Even slight overtreatment of water can result in posttreatment precipitation within the distribution system of deposits containing alum, lime, or calcium carbonate. A utility may promote controlled precipitation to lay down a thin layer (eggshell coating) of calcium carbonate for protection of the metallic pipeline interior. However, excessive or irregular deposits can easily occur, requiring cleaning of the distribution system.

Encrustation

Encrustation is a by-product of corrosion (tubercles) mixed with mineral deposits, such as iron, manganese, and carbonates. Before the 1960s, many iron pipes were

Figure 1-1 Pipe with problems caused by corrosion

installed without linings to protect the interior surfaces. These unlined pipes often experience internal corrosion.

As corrosion occurs, the interior of the pipe develops pits from which material is removed and tubercles where material is deposited (Figure 1-1). Additionally, corrosion can create "red water" complaints from end users. Corrosion may result from direct oxidation or electrolytic action, both fostered by aggressive water. Tuberculation can vary with water chemistry (from very soft to very hard water). Proper water chemistry does help to prevent additional buildup of encrustation. Most encrustation can be removed by cleaning. Perhaps the most significant water quality problem presented by encrustation are the potential difficulties maintaining a disinfectant residual. Removal of encrustation often increases a system's disinfectant residual; however, regrowth of encrustation is likely in unlined iron pipes.

Fouling

Fouling represents a very significant problem, but one that is not always well-understood. A fouling problem can develop with any type of pipe material. The condition is usually caused by natural biological activity and results in buildup of an organic deposit on the interior of the pipe. Although this deposit is often soft and filamentous, it can severely affect water turbidity and cause taste-and-odor problems. Bacteriological activity from organisms, such as iron-fixing bacteria, can result in development of slimes and severe deposits in the pipe.

HYDRAULIC PERFORMANCE

In addition to water quality problems discussed so far, any buildup inside distribution system piping can greatly reduce the hydraulic performance of the system. Hydraulic engineers have long determined flow in pipes by using the empirically derived Hazen-Williams formula:

$$V = 1.318 C r^{0.63} s^{0.54}$$

Table 1-1   Hazen-Williams coefficient

Condition	C
New pipe	130–140
Fair to normal (interior clean)	100
Significant reduction in pipe capacity	70
Severe problem—interior cross section greatly reduced	30–50

where:

V = velocity, in ft/sec (m/sec)

r = hydraulic radius, in ft (m), which is the cross-sectional area of the pipe divided by the wetted perimeter

C = Hazen-Williams roughness coefficient

s = slope of the hydraulic grade line, in ft/ft (m/m)

C (often called the "C factor") represents the coefficient of friction, a measure of the roughness of the interior of the pipe. Expressed in terms of C, the formula can be stated as:

$$C = 2{,}466QD^{-2.63}H^{-0.54}L^{0.54}$$

where:

C = Hazen-Williams roughness coefficient

Q = Quantity of flow in a pressure conduit, in mgd (m³/d)

D = Nominal diameter of the pipe, in in. (mm)

H = Head loss, in ft (m) of water

L = Length of pipe, in ft (m)

The C factor, and hence the flow in a pipeline, depends on the type of pipe and its interior condition (see Table 1-1). For a given velocity, increased internal surface roughness (changing laminar to turbulent flow) can lead to a reduction in overall pipeline efficiency.

Field testing techniques allow distribution system operators to calculate Hazen-Williams C factors for their systems. These data help in making informed decisions about which process to employ to restore hydraulic efficiency. Collecting data for the Hazen-Williams C factor after employing any cleaning or pipe rehabilitation process is also a very useful way to gauge the impact of the system improvements.

STRUCTURAL PERFORMANCE AND LEAKAGE

The structural performance of a water main may deteriorate over time because of a number of effects. Cast-iron, ductile-iron, and steel piping may be subject to internal and/or external corrosion, resulting in pitting and wall thinning, which can lead to leakage and eventual burst failures. Cement-based pipes such as asbestos–cement and concrete piping may also be subject to deterioration due to corrosion of the cement matrix and/or steel reinforcement. In addition, all types of pipe, including thermoplastics, may be subject to joint failure between pipe lengths and hence excessive leakage, which can in turn lead to washout of bedding and subsequent structural failure.

Such structural and leakage failures can have direct consequences such as high maintenance costs, water quality problems, service interruptions, and loss of valuable treated water. They may also have indirect consequences in terms of the economic and public relations costs of damage associated with bursts and the overall public image of the service provider.

REHABILITATION SOLUTIONS

This manual describes a number of possible solutions to problems arising from corrosion and deposition. These range from simple periodic cleaning to replacement of the pipe using "trenchless" techniques. All of the solutions discussed in the manual make some use of the existing pipe, either as part of the rehabilitated system (renovation solutions) or as a convenient route for installation of new piping (replacement solutions). Solutions involving installation of a replacement pipe along a new route, such as open trench laying, directional drilling, and microtunneling, are outside the scope of this manual.

Selecting the optimal solution to a specific pipeline problem is a complex process involving both technical and economic considerations. Both the American Water Works Association Reseach Foundation and a number of AWWA technical committees are developing computer-based decision tools to assist utility engineers in this process. This work is expected to come to fruition while this edition of the manual remains in effect. In the meantime, the following guidelines may prove useful.

SELECTION OF REHABILITATION SOLUTIONS

Key elements in the selection of a rehabilitation solution are

1. The exact nature of the problem(s) to be solved

2. The hydraulic and operating pressure requirements for the rehabilitated main

3. The materials, dimensions, and geometry of the water main

4. The types and locations of valves, fittings, and service connections

5. The length of time in which the main can be taken out of service

6. Site-specific factors

The aim of the selection process is to consider all these factors to arrive at the most cost-effective, technically viable solution. Ideally, the cost estimate should include not only direct contracting and related costs but also indirect costs associated with public disruption and longer-term maintenance and other "life cycle" costs.

One approach to technique selection is summarized in Figures 1-2, 1-3, and 1-4. Together, these charts provide a framework for selecting or rejecting groups of techniques, depending on the nature of the performance problems, hydraulic requirements, and some site-specific factors. In some cases, the charts indicate use of lining techniques classified as either Class I (nonstructural), Class II/III (semistructural), or Class IV (structural). A more detailed discussion of this classification system and of other key design issues associated with such lining techniques is presented in appendix A.

The figures do not list cleaning as a solution for water quality or flow and pressure problems. Cleaning with one of the various techniques discussed in the manual may well offer the lowest-cost immediate solution to many of these problems. It may offer a long-term solution if repeated as required or combined with chemical treatment of water to prevent or delay recurrence of the original problem. However, cleaning is more frequently used as a necessary preliminary step before carrying out one of the lining processes described in the manual.

DISTRIBUTION SYSTEM PERFORMANCE CRITERIA 5

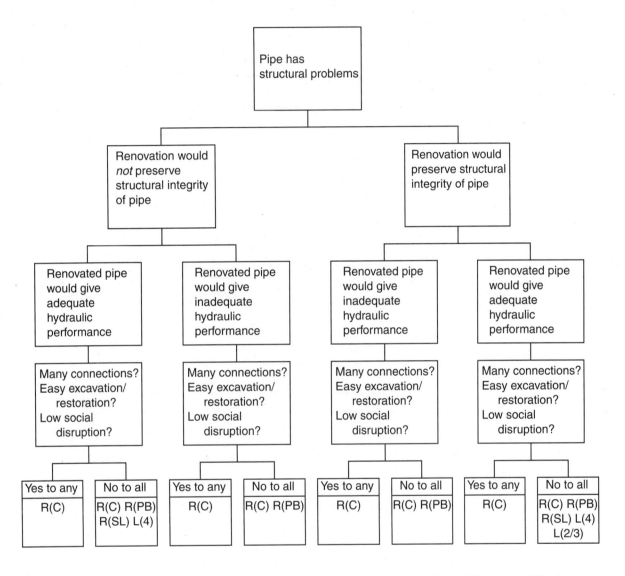

R(C)–Replacement (conventional or boring/directional drilling)
R(PB)–Replacement (pipe bursting)
R(SL)–Replacement (slip-lining)
L(2/3)–Lining (semistructural—Class II/III)
L(4)–Lining (structural–Class IV)

Figure 1-2 Selection of rehabilitation techniques to resolve structural problems

6 REHABILITATION OF WATER MAINS

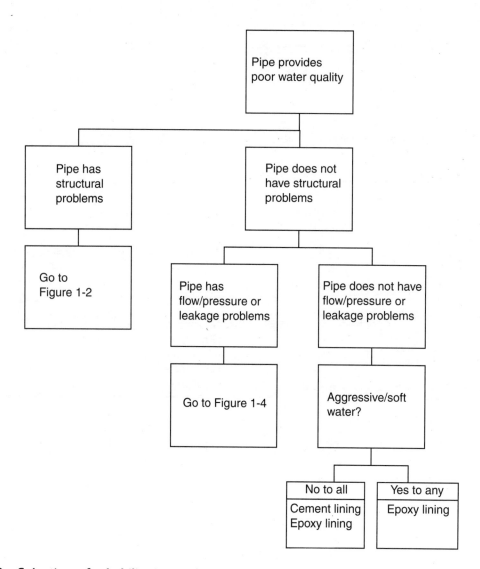

Figure 1-3 Selection of rehabilitation techniques to resolve water quality problems

DISTRIBUTION SYSTEM PERFORMANCE CRITERIA 7

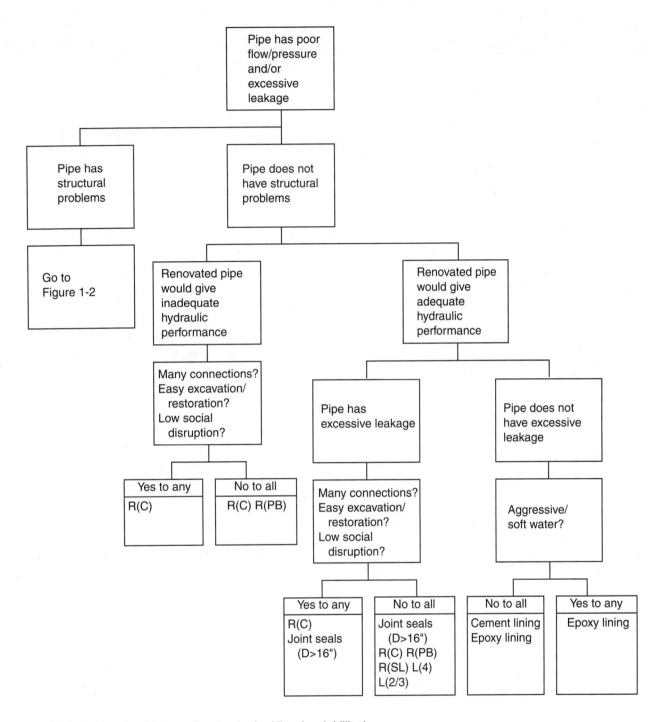

R(C)–Replacement (conventional or boring/directional drilling)
R(PB)–Replacement (pipe bursting)
R(SL)–Replacement (slip-lining)
L(2/3)–Lining (semistructural—Class II/III)
L(4)–Lining (structural—Class IV)

Figure 1-4 Selection of rehabilitation techniques to resolve flow, pressure, and leakage problems

This page intentionally blank.

AWWA MANUAL M28

Chapter 2

Cleaning

Water mains can be cleaned in place in a water distribution system using one or more of the following methods:

- Flushing
- Cable-attached devices
- Fluid-propelled devices
- Power boring

Dispose of cleaning water and debris in accordance with applicable regulations. Refer to AWWA Standard C651, Disinfecting Water Mains, for guidance in disinfecting pipelines before returning them to service.

FLUSHING

Water main flushing has a definite place in distribution system operation. It is particularly effective for light cleaning jobs. Mains should be flushed to clean newly installed and repaired mains prior to and after disinfection. Routine flushing is needed to remove impurities that cause complaints or are considered hazardous to public health. Refer to the AWWA handbook titled *Maintaining Distribution-System Water Quality* for specific flushing program guidelines.

Flushing should never be considered the only solution to water quality problems identified in the distribution system. Proper design and operation of distribution facilities and an effective backflow-prevention program should also be part of any effort to maintain water quality. While quality control is the primary purpose of flushing, careful observation of system hydraulics during flushing may indicate problems in mains, such as inadequate capacity, undiscovered restrictions, or closed or partially closed valves.

10 REHABILITATION OF WATER MAINS

Figure 2-1   Drag cleaning, in which a winch pulls a mechanical cleaner through the pipe

CABLE-ATTACHED DEVICES

Systems that use cable-attached devices for cleaning distribution mains include drag cleaning, hydraulic-jet cleaning, and electric scraper cleaning. In each case, the length of hose or cable determines the length of the pipe section that can be cleaned.

Drag Cleaning

In drag cleaning, a winch pulls a mechanical cleaner, composed of a series of steel scraper blades and rubber squeegees, through the pipe (Figure 2-1). The mechanical cleaner is usually flexible, allowing it to negotiate bends of up to 45°. Both ends of the cleaner are fastened to steel cables, which are attached to winches at either end of the pipe section to be cleaned. The cleaner is winched first in one direction and then in the other, and this process continues until the pipe is satisfactorily cleaned.

This cleaning method offers important advantages:

- Cleaning can be accomplished when water pressure or volume is insufficient to propel a hydraulically driven device, or when excessive pressure would be required for hydraulic cleaning, especially with small-diameter mains.

- Very hard deposits and encrustation can be removed.

- Dry solids allow easy disposal.

Hydraulic-Jet Cleaning

To clean pipe with a hydraulic jet, a special nozzle attached to a hose emits a jet of water at high velocity and pressure that removes debris and deposits from the interior of the pipe. Nozzle pressures of 1,000 to 10,000 psi (6,900 to 69,000 kPa) and above can be generated using this system. The jetted stream of water must bear hard enough against the scale or deposit to breach it and dislodge particles. Once the scale or depositional material is penetrated, the fluid forms a wedge between the deposit and the surface and strips off the deposit, exposing the clean metal surface. In many jetting operations, dislodged particles are entrained in the jetting stream and effectively dislodge more particles. The principal advantage of this method is removal of very tough deposits.

Electric Scrapers

Electric scrapers are used to clean large-diameter lines. A variation of this cleaning method is the use of an electrically powered, operator-driven scraper. This power-driven scraper incorporates revolving brushes or rotating arms to clean the line.

The principal advantage of this method is the ability of the operator to evaluate the effectiveness of the cleaning process as it proceeds through the line.

FLUID-PROPELLED CLEANING DEVICES

This section of the chapter describes fluid-propelled cleaning devices such as foam pigs and mechanical metal scrapers.

Foam Pigs

Foam pigs are flexible, bullet-shaped cleaning tools manufactured of high-quality (5 to 8 lb/ft^3 [80 to 128 kg/m^3] density) open-cell polyurethane foam, some coated with polyurethane synthetic rubber coatings (70 lb/ft^3 [1,121 kg/m^3] density). (See Figures 2-2 and 2-3.) Pigs are propelled down water mains by the pressure and volume of water in the distribution system. Cleaning is accomplished by the frictional drag and flexible characteristics of the foam pig, which removes foreign objects, iron tuberculation, and other matter as it passes through the pipes, leaving the interior surfaces as smooth and free from irregularities as age, attrition of use, and type of piping material will allow. When water pressure is applied for propulsion, a certain amount of water bypass (approximately 10 percent) helps to keep loose debris suspended out in front of the foam pig (Figure 2-4).

Cleaning of deteriorating mains generally requires a series of swabs and foam pigs applied in progressively larger diameters (referred to as *progressive pigging*) until the pipe is restored to its original diameter.

Figure 2-2 Foam pig: A bullet-shaped device made of polyurethane foam

Figure 2-3 A foam pig with hardened coatings

12   REHABILITATION OF WATER MAINS

Figure 2-4   Loose debris flushed ahead of the pig

Operating procedures. The following steps briefly describe the most common operating procedures for cleaning water distribution mains with foam pigs. The described procedure is based on a best-case scenario; actual procedures may vary for certain applications.

1. Review drawings of lines to be cleaned to identify a) possible entry and exit points, and b) all valves to be used to isolate the section of system to be cleaned.

2. Exercise all valves in the section to be cleaned before the pigging operation to ensure isolation of the section to be cleaned.

3. Flow tests should be performed before and after the pigging operation to evaluate the efficacy of the cleaning and determine the resulting condition of the water main.

4. Provide adequate means for disposing any debris exiting the pipe. Storm drains may be suitable for disposing of small amounts of light debris. However, the best arrangement for handling large amounts of heavy debris would be to direct the flow to a temporary settling pond, the bed of a dump truck, etc., to facilitate removal of solids once the water has drained away.

5. Notify all affected customers (commercial and residential) as well as fire department officials of the scheduled interruption of service.

6. Confirm that all valves in the section to be cleaned are fully opened and working properly and that the section is properly isolated. Foam pigs follow the direction of flow, which must be controlled by the valve locations.

7. Introducing the foam pigs into the water main may be accomplished either by hand or through the use of mechanical equipment commonly called a *pig launcher*. Several types of fire hydrants, once they are disassembled, can be used as entry and exit points for pigging of some water mains 6 in. (150 mm) and smaller (Figure 2-5).

8. To produce the best cleaning results, foam pigs should travel through a water main at a rate of 3 to 9 ft/sec (9 to 24 m/sec).

9. Sequence of foam pig runs:

    a. Introduce a line-sized, soft-foam swab (1.5 to 2.0 lb/ft^3 [24 to 32 kg/m^3] density, uncoated), called a *prover pig*, to determine the actual effective inside diameter of the pipe to be cleaned. The swab wears down while passing deposits in and along the pipe wall.

    b. Introduce a line-sized bare foam pig (5 to 8 lb/ft^3 [80 to 128 kg/m^3] density, uncoated) to remove soft deposits and help gauge the true opening in the line.

    c. Introduce a line-sized, coated foam pig (5 to 8 lb/ft^3 [80 to 128 kg/m^3] density, coated with urethane elastomer). This type of foam pig should be run repeatedly until one emerges in reusable condition. A water line with excessive buildup should be pigged using the progressive pigging method, beginning with undersized, coated foam pigs and gradually increasing the size in successive passes until a line-sized foam pig emerges in reusable condition.

Figure 2-5 Pigs launched through a disassembled fire hydrant for a 6-in. (150-mm) or smaller line

14 REHABILITATION OF WATER MAINS

d. Line-sized, wire brush foam pigs may be introduced following these applications for final removal of extremely hard deposits (such as tuberculation).

e. Introduce a line-sized swab (1.5 to 2.0 lb/ft^3 [24 to 32 kg/m^3] density foam, uncoated) to sweep out any loose debris and determine the effectiveness of the cleaning process.

10. Flush and disinfect as required.

11. Place the water main back into service.

Foam pigs are flexible enough to negotiate short-radius bends and pass through full-port valves and some types of plug valves. Their capability to be compressed up to 35 percent of their cross-sectional area allows them to pass through multidimensional lines and full-port valves. A piggable valve is one through which a pig will pass every time. Foam pigs do not pass through butterfly valves. Foam pigs can clean long sections of piping from one entry point, thus reducing the number of excavation sites required. Pigging is a fast, simple, and cost-effective method of returning most deteriorated piping systems to as-designed capacity and efficiency.

Metal Scrapers

A metal cleaning scraper consists of a steel frame shaped like a piston. Specially tempered steel blades are attached around the scraper at various angles to create a scraping and brushing action (Figures 2-6 through 2-8). The cleaner is propelled through the water main by means of water pressure.

Cleaning is often accomplished with a single pass in a continuous operation; sometimes, however, interior pipe conditions may require additional passes. The length of pipeline that can be cleaned hydraulically in one operation is limited only by the availability of volume and pressure water and a proper means of disposing of water and deposits. An opening must be provided at each end of the section to be cleaned for entry and exit of the cleaning tool. The volume of water required to hydraulically clean

Figure 2-6 Scraper unit with specially tempered steel blades

Figure 2-7 Several scraper units assembled together in the field

Figure 2-8   A series of disks to act as a hydraulic piston, pulling scrapers through the line

a pipeline will depend, to a great extent, on how dirty the water is. Sufficient water must be added behind the cleaner to fill the pipe as the cleaner moves ahead.

The water that passes the cleaner scours the wall of the pipe and washes ahead the material that is scraped off the pipe. While the velocity of water ahead of the cleaner is independent of cleaner speed, it must be sufficient to remove the deposits. Experience indicates that a flow velocity ahead of the cleaner of between 2 and 10 ft/sec (0.6 and 3.0 m/sec) is required to remove the deposits. In a small pipe, because a relatively large amount of material must be moved for the volume of the pipe, a relatively high velocity is required.

The cleaning water and deposits must be discharged from the pipeline in a way that avoids creating a nuisance or environmental problem. A sandbag dam can be constructed to create a pond for particle settlement (Figure 2-9). This allows the clean water to be decanted to a storm drain while any solid material is collected and properly disposed of. Another method is to pipe the water to a dump truck, catching the solids in the truck and allowing the water to run off into a storm drain.

Operating procedures. The following steps describe the operating procedures commonly used when cleaning water pipes with mechanical scrapers.

1. Review drawings of lines to be cleaned. Identify logical entry and exit points for the cleaning tool. Note all valves, service connections, and other components that must be opened or closed to isolate the line to be cleaned. Conduct C-factor tests (if not previously completed) to determine the condition of the line.

2. Excavate and install a spool piece at each entry and exit point (Figure 2-10). Obtain a sample of deposits in the line and determine their average thickness.

3. Determine source of cleaning water (e.g., reservoir, feeder mains, parallel lines, etc.). Cleaning for lines 12 in. (300 mm) in diameter and smaller may require auxiliary pumps to provide adequate flow and pressure; flow and pressure within the system generally are adequate to clean mains over 12 in. (300 mm) in diameter.

4. Provide suitable means for disposing water and removed solids.

5. Notify all affected commercial and residential customers, as well as fire department officials, of the scheduled interruption of service.

16 REHABILITATION OF WATER MAINS

Figure 2-9 A sandbag dam to create a pond for particle settling

Figure 2-10 A spool piece installed at the entry and exit points for mechanical scrapers

6. Completely isolate the section to be cleaned. Close valves slowly to prevent water hammer.

7. Drain the line to be cleaned.

8. Open entry and exit points and remove spool piece sections. At the entry point, install a spool piece with the cleaning device inside. At the exit point, install discharge piping to bring discharged water up to ground level. Connect a water manifold with a throttle valve to control the rate of travel of the cleaning device.

9. Initiate the required flow into the main, and control speed with a bypass, blowoff valve.

10. Complete the cleaning operation.

11. Reclaim the cleaning device at the exit spool.

12. Repeat additional cleaning passes, as required.

13. Remove the insertion and retrieval sections, and replace permanent spool sections. Backfill and complete other necessary maintenance.

14. Flush and disinfect the line, as required. Conduct tests to determine its C factor following cleaning.

15. Return the line to service.

Advantages. Principal advantages of this method include

- Ability to clean long stretches of heavily deposited pipe at 2 to 10 ft/sec (0.6 to 3.0 m/sec) in a single, continuous operation

- Flexibility to negotiate standard bends (with maximum radius equal to 1½ pipe diameters) and elbows, inclined and vertical pipe sections, and to pass line-size gate valves, ball valves, tees, and corporation taps

- Ability to clean water mains with minimal excavation points

- Capability to restore water mains to C factors comparable to newly installed, unlined pipe

CLEANING BY POWER BORING

Power boring is a cleaning method using any hydraulically powered device capable of removing tuberculation and encrustation from cast-iron, ductile-iron, and steel pipe from 3 in. (76 mm) in diameter and above. This process is normally carried out in lengths of 400 ft (121 m) or greater. This cleaning method is described in the following paragraphs.

Rack-Feed Boring Equipment

A rack-feed boring machine (Figure 2-11) is a compact, diesel-powered unit that uses hydraulic pressure to deliver up to 31 horsepower (23.1 kW) to a boring head. The boring head is designed to accommodate spring steel boring rods 15 ft (4.6 m) long fitted with spring-loaded quick-connects for connecting rods into suitable lengths for cleaning various lengths of pipe (Figure 2-12). The end of a boring rod assembly is fitted with a spring steel cutter blade or other cleaning tool, which rotates at 750 rpm through the pipe. (See Figure 2-13.) This cleaning process is conducted against a controlled, upstream water flow to flush loosened debris from the pipe.

18 REHABILITATION OF WATER MAINS

Source: Courtesy of AdTec International, Inc. (Gary Zinn), and Mercol Products Ltd.

Figure 2-11 Rack-feed boring machine

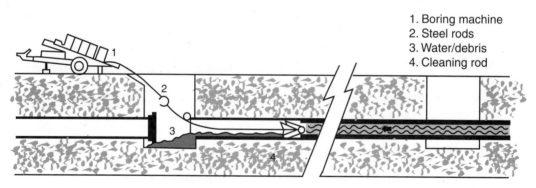

1. Boring machine
2. Steel rods
3. Water/debris
4. Cleaning rod

Source: Courtesy of AdTec International, Inc. (Gary Zinn), and Mercol Products Ltd.

Figure 2-12 Cleaning pipe by power boring

    The rack-feed boring machine may be equipped with an adjustable boom to accommodate various pipe depths and to control the angle at which boring rods are inserted into the pipe. The ratio of boring rate to spring cutter blade revolutions is predetermined and fixed to eliminate operator error. This setting ensures consistent results throughout the cleaning operation.

    The rack-feed boring method leaves a pipe's interior surface free from tuberculation and encrustation and can be effective for bends up to 22.5°. Bends of greater radius may require removal and replacement.

CLEANING 19

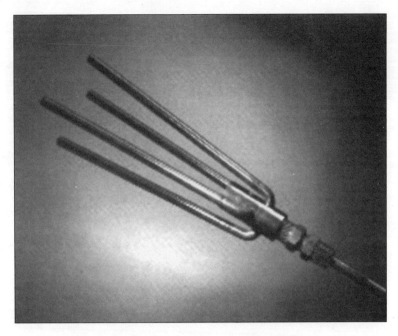

Source: Courtesy of AdTec International, Inc. (Gary Zinn), and Mercol Products Ltd.

Figure 2-13 Cleaning head

Operating procedures. The following steps briefly describe the basic operating procedures:

1. Review the drawings of lines to be cleaned and verify by site inspection all possible entry and exit points and the locations of valves, fittings, etc. Use pipe-locating equipment and clearly mark buried utility services.

2. Coordinate any remedial action with the utility.

3. Conduct flow tests (if required) before and after cleaning to determine the condition of the main.

4. Notify all customers of work to be conducted.

5. Lay temporary service, chlorinated in accordance with established standards.

6. Access all points of entry in accordance with contractual documents.

7. Following the boring machine manufacturer's operating instructions, bore clean the pipe. Capture all debris for proper disposal.

8. Following bore cleaning, remove residual water from the pipe using foam swabs and/or rubber squeegees.

9. If available, color closed-circuit television equipment may be used to inspect the pipe and produce a video, noting street names and addresses inspected.

This page intentionally blank.

AWWA MANUAL M28

Chapter 3

Lining Techniques

A water utility may decide to line its distribution system pipes in response to a number of problems, such as those described in chapter 1. Lining a pipeline can eliminate the need for frequent flushing. Where leakage is a problem, linings have been installed to reduce or eliminate leaks through corroded areas of pipe or bad joints. A smooth lining in a corroding pipe helps to maximize hydraulic carrying capacity and minimize pumping costs. Additionally, some lining systems can correct structural failures, bridge breaks, and missing sections in corroded pipe, thus restoring service through a continuous pipeline.

In-place lining of water mains can be accomplished by one or more of the following general methods:

- Cement–mortar lining
- Epoxy lining
- Slip-lining
- Modified slip-lining
- Cured-in-place lining
- Structural lining
- Internal joint seals
- Pipe bursting

Internal joint seals and pipe bursting methods are discussed in chapters 4 and 5.

Refer to AWWA C651, Disinfecting Water Mains, for guidance on procedures for disinfecting pipelines before returning them to service.

CEMENT–MORTAR LINING

When alkaline-containing water comes in contact with iron, a chemical inhibitor against oxidation forms. Because the cement and sand in cement–mortar lining is porous, water can penetrate through the lining to the pipe wall, becoming alkaline in the process. Consequently, pipe that is lined with cement mortar is protected from oxidation because of the composition of the portland cement.

Cement–mortar linings were first installed in existing pipelines using the centrifugal process in the mid-1930s. However, this method was limited to pipelines large enough for a person to enter. In the 1960s, remote lining processes were introduced. Today, cement mortar is applied to new ductile-iron pipes and most new steel pipes before installation, making this method a standard in the water industry.

Cement mortar is applied to the pipe wall by the rotating head of an electric or air-powered machine. The machine is placed in the pipe at 300 to 1,500 ft (92 to 458 m) intervals, depending on pipe diameter, valve locations, bends, profile, and alignment (Figures 3-1, 3-2, and 3-3). Mortar is pumped to the lining machine through high-pressure hoses or is mechanically delivered. The lining machine is equipped with rotating trowels or a conical drag trowel positioned just behind the dispensing head. As the machine moves through the pipe, it leaves a smooth, troweled (nonstructural) finish.

A reinforced cement–mortar lining may also provide structural improvement. Initially a 0.5-in. (13-mm) thick cement–mortar lining is placed on the pipe wall in the conventional manner without troweling. Using overlapping joints, a wire mesh is placed against the lining, which is then covered by another 0.5-in. (13-mm) troweled cement–mortar lining. The remaining steps in the process are the same as those for an unreinforced lining. This process is generally used to rehabilitate steel lines in poor condition due to extensive corrosion and in lines large enough for a person to enter. The reinforcing wire holds the cement mortar together, even if large holes develop in the pipeline. While this method will not necessarily prevent a blowout, it will extend the useful life of a pipeline.

Operating Procedures

The following steps describe the common operating procedures used when lining a main with nonstructural cement mortar.

Figure 3-1   A cement–mortar lining machine for use in small-diameter pipe

LINING TECHNIQUES 23

Figure 3-2 Introduction of a small lining machine

Figure 3-3 A cement–mortar lining machine for use in large-diameter pipe

1. Topography, bends, in-line valves, reduced-diameter in-line valves, and traffic conditions all influence the locations of excavations for access points to allow introduction of the lining equipment. These conditions also affect the method used to clean the pipe. All 22.5°, 45°, and 90° bends must be removed for 12-in. (300-mm) diameter pipe and smaller; 45° and 90° bends must be removed in 16-in. (400-mm) diameter pipe. Pipe 20-in. (500-mm) diameter and larger may require excavations near the bends or removal.

   Full-diameter line valves must be removed and replaced, or the bonnets removed to clean the cement mortar from the interior of the valve if the pipe is too small for a person to enter. In large pipe, personnel can clean the valve interior from inside the pipe. Depending on their size, reduced-diameter or obstructed line valves may require removal or excavations on each side of the valves to allow lining.

2. The pipeline is generally dewatered by gravity and/or blowoffs, which remove most of the water. The contractor can then remove any remaining water lying in low spots with rubber squeegees pulled by winch through the pipe. At this time, inspection usually determines whether any branch valves are leaking. If a valve fails to stop water from entering the pipe to be lined, it is reexercised. If the leakage continues, a seal is installed or the valve is replaced. The pipe must be free of water when lining with cement mortar.

3. Cast-iron pipe is cut with a squeeze-type cutter, guillotine saw, handheld mechanical saw, or traveling saw, and a 4- to 5-ft (1.0- to 1.5-m) section of pipe is removed. Steel pipe is cut by torch, and a half cap is removed. Mechanical couplings are used for reconnecting cast-iron pipe, and butt straps are used to reconnect steel pipe.

4. A mixing van or concrete batch plant is located near the access hole where a cement–mortar mixture, consisting of one part silica-type sand and one part type II portland cement, is prepared. The mixed mortar is then delivered to the lining machine by one of several methods, depending on the pipe diameter.

5. In 4- to 24-in. (100- to 600-mm) diameter pipe (sometimes larger), the mortar is pumped through high-pressure hoses to the lining machine. Specially designed winches pull the lining machine through the pipe at a constant speed, ensuring a uniform lining thickness.

6. Mechanical feeding equipment shuttling between the access excavation (where the mortar is mixed) and the lining machine delivers mortar to the lining-machine hopper. The lining-machine operator then regulates the mortar application. Mechanically driven rotating trowels are used with this manually operated equipment.

7. Consumers' service lines and laterals less than 2 in. (50 mm) in diameter must be cleared after the lining application. This is done about 1 hr after the lining is completed, using compressed air to blow open the service line at the connection to the main. Laterals over 2 in. (50 mm) in diameter are not plugged by centrifugal lining and do not require excavation or blow back.

8. Water can be introduced into the line without pressure to allow curing 24 hr after completing the lining. The main can then be chlorinated, tested, and

Figure 3-4 A pipe ready to be returned to service four to seven days after cement–mortar lining

returned to service. (Discharge water should be disposed of in accordance with local ordinances. Chlorine may need to be neutralized and the pH adjusted before discharge. Mains with low flow may experience high pH problems for a short time after being returned to service.) Most distribution mains are returned to service between four and seven days after lining (Figure 3-4), depending on valve locations and disinfection requirements.

9. AWWA C602, Standard for Cement–Mortar Lining of Water Pipelines in Place—4 in. (100 mm) and Larger, covers cement–mortar lining of pipelines from 4 in. (100 mm) to 144 in. (3,658 mm) in diameter.

EPOXY LINING

The process for in-situ epoxy resin relining (ERL) of iron and steel pipelines was developed in the United Kingdom in the late 1970s, and has been performed in North America since the early 1990s. The process has been used effectively to rehabilitate old, unlined water mains. The epoxy materials approved for use were first certified by ANSI/NSF Standard 61 in 1995.

Epoxy lining of potable water mains is currently classified as a nonstructural renewal method. The process involves cleaning the pipe to remove existing corrosion buildup and then spraying a thin (1 mm) liquid epoxy coating onto the inner wall of the pipe. The coating cures in 16 hr and provides a smooth and durable finish resistant to mineral deposits and future turberculation buildup.

Prior to lining, pipes must be thoroughly cleaned to remove tuberculation and produce a clean surface to which the epoxy lining will adhere. Either the power-boring or drag-scraping techniques, described in chapter 2, will provide a sufficiently clean surface for ERL. The pipes are then dried by plunging or swabbing prior to lining.

26 REHABILITATION OF WATER MAINS

Figure 3-5 A typical epoxy lining machine

Several epoxy lining materials are currently approved for use in potable water systems under ANSI/NSF Standard 61 guidelines. Epoxy resin products are two-component systems classified as 100 percent solids by volume with no reactive chemical agents present. Epoxy linings are applied with specially designed machines of several different types (Figure 3-5). Most (but not all) have separate, heated reservoirs from which positive-displacement pumps precisely control the quantities of resin and hardener applied to the pipes.

Most lining machine models are computer controlled with warning devices that alert operators if the minimum lining thickness is not being achieved. A lining machine applies the epoxy material with an application head attached to the lining hoses (Figure 3-6). This head consists of an in-line static mixer followed by a centrifugal spinner applicator.

The applicator head and hoses are pulled to the far end of the cleaned pipe length and then winched back through the pipe at a speed linked to the rate of supply of the epoxy resin mixture. In this way, a coating with a typical uniform thickness of 1 mm (minimum) is applied.

After lining, the ends of the pipe are capped, and the resin is allowed to cure overnight at ambient temperature. The pipe then is flushed, disinfected, and returned to service (Figure 3-7). The following section discusses procedures and major elements of a successful epoxy lining project.

LINING TECHNIQUES 27

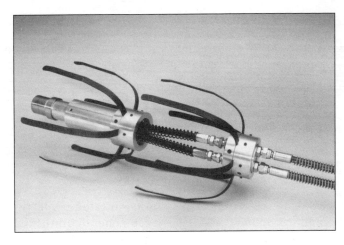

Figure 3-6 A typical epoxy lining application head

Figure 3-7 An epoxy-lined water main

Operating Procedures

1. Locate excavation points, and evaluate in-line bends, valves, and other conditions to determine access locations.

2. Thoroughly clean pipes, remove standing water, and inspect before lining operations begin.

3. Prior to lining, conduct checks on the equipment. Verify pump output, mix ratio, and material temperatures. Record this and other pertinent site information on a lining operation record sheet. Complete a separate record sheet for each separate lining run.

4. Prior to inserting delivery hoses into the main, the operator should pump and recirculate the epoxy components until the material reaches the uniform operating temperature specified by the epoxy manufacturer.

5. Once the hoses are inserted into the pipe and the static mixer and application head are attached, check for proper operation. Visually check for correct mixing of the two epoxy components by test spraying the mixed epoxy into a container or onto a test card and recording the observed epoxy color on the record sheet.

6. The application can begin when the operator is satisfied that the material flows are established and the lining material color is consistent.

7. The winch speed and the rate of withdrawal of the hoses should be carefully monitored to ensure a) a smooth traverse, and b) that the rate of withdrawal leaves a minimum lining thickness of 1 mm (40 mil) in a single application.

8. Epoxy resin lining should not be performed when the pipe temperature is below 38°F (3°C).

9. Immediately after completing application of the epoxy lining and inspecting the section of pipe, cap the ends of the main to prevent contamination and/or water from entering the pipe.

10. Allow the minimum cure period specified by the epoxy manufacturer (usually 16 hr).

11. After the cure period, visually inspect the pipe at both ends. Complete closed-circuit television inspection is recommended until the utility is satisfied with the quality of application from the contractor or operators.

SLIP-LINING

Another viable method of rehabilitating existing water pipelines is the insertion of flexible thermoplastic liners directly into the mains. This process, commonly known as *slip-lining*, has been widely used by sewer and natural gas utilities since the early 1980s. However, its use in transmission and distribution systems for source water and treated, potable water has not been as widespread as other pipe-lining techniques. The key benefit of slip-lining is that it creates a new, integral pressure pipe inside the old, deficient pipeline without a complete excavation.

Using a process known as thermal butt fusion, the ends of several consecutive 40-ft (12-m) lengths of flexible pipe are joined at a convenient location aboveground to form a single length of pipe, perhaps hundreds of feet (100 m or more) long. One end of this pipe is then pulled by cable into the entry pit and through the section of old pipe. The new pipe is then reconnected to the existing mains.

Potential applications for slip-lining are numerous. Most existing pipelines can be slip-lined, but certain applications are ideally suited to this method:

- Where poor structural integrity of existing pipes makes other lining methods, such as cement mortar, inadvisable

- When service connections and branches are limited

- Where a structure has been built over the existing main, making replacement economically impractical

- Where a main crosses over or under railroads, bridges, rivers, or other obstacles, making alternative linings impractical or not economically feasible

- Where other unique circumstances make alternative lining methods impractical

An inserted liner does substantially reduce the effective cross-sectional area of the pipe. Consequently, postlining flow requirements must be considered when deciding to slip-line. However, the reduction in the friction factor of the liner pipe as compared to the old, unlined pipe should significantly compensate for the reduced internal diameter. In addition, the flow rate will not be reduced by corrosion or scale over time, as would be expected of many other unlined piping materials. Finally, the geometry of the unlined pipe must be considered, as liners generally do not turn well through elbows.

Operating Procedures

The following steps describe the common operating procedures for slip-lining a main:

1. In general, the insertion pipe should be sized so that its outside diameter is at least 10 percent smaller than the inside diameter of the pipe being lined to allow for smooth insertion. Possible obstructions at pipe joints and taps, and the normal friction created during the insertion process dictate a conservative approach to liner pipe sizing. Pipe manufacturers recommend sizing of available liners. Most pipe sizes are standard iron pipe size, but special diameters are also available for slip-lining.

2. The type of material for the slip-lining pipe should be carefully chosen. The most common material specified is thermoplastic pipe manufactured of high-density polyethylene (HDPE). This material can handle moderately high temperatures and highly corrosive and abrasive liquids in nearly all applications. For some special applications, the use of other thermoplastics such as polypropylene or polybutylene may be recommended.

   Factors to consider in material selection include wall thickness needed to achieve the desired pressure rating at the design temperature and ability to withstand thermal expansion, soil loads, and traffic loads. A pipe's ability to withstand the rigors of insertion is generally related to the use of proper installation procedures and not the pipe material, since even the thinnest commercially available HDPE pipes are routinely slip-lined in lengths of several hundred feet (100 m or more).

   The pressure rating is determined by the type of piping materials[*] and the standard dimension ratio (SDR), which is the ratio of the outside diameter to the minimum wall thickness. All diameters of pipe of a single

---

[*]Examples are PE3406 and PE3408, both designations of the Plastic Pipe Institute, a division of the Society of the Plastics Industry, Inc., 355 Lexington Ave., New York, NY 10017.

type having a given SDR value, such as SDR-11, will have the same pressure rating and can be combined into one piping system. Any slip-lined pipe is fully pressure rated.

A variety of special fittings, taps, and other appurtenances are manufactured to make transitions between HDPE pipe and old cast-iron and steel mains and to reconnect customers and branches to the newly inserted plastic water main.

3. Prior to slip-lining a main, detailed plans should be developed of the existing pipelines, valves, branch pipelines, service connections, and fire hydrants. Because slip-lining sometimes requires large excavations, special attention must be given to traffic control and public safety. Because the entire existing pipeline is not excavated, locate excavations away from highly traveled areas whenever possible. Plans and specifications should clearly spell out required standards of line installation and related work.

4. Pipelines for customer service connections or for fire protection may have to be bypassed. Although downtime required for slip-lining can be relatively short, the need for temporary service should be considered.

5. The designer must decide what work other than actual slip-lining will be performed. Directly related to the lining process, of course, is the reconnection of services, fire hydrants, and sideline laterals, and connection of the newly relined system to adjacent sections of the water system. Other work could include replacement of sections and in-line valves, service piping, and corporation cock replacement. Also, fire hydrant and sideline laterals may be slip-lined or replaced, and lateral valves and fire hydrants may be replaced.

6. If temporary service is to be provided, it should be installed before beginning the insertion process.

7. A minimum of two excavations are normally required for slip-lining, one at each end of the segment of pipe to be slip-lined. Commonly, excavations at elbows allow equipment to pull in two directions from one location. The excavation at the insertion end should have a minimum slope of 3 ft per foot of depth (3 m per metre of depth), plus additional trench length and breakout above the spring line of the old pipe of 1 ft/in. (1 m/80 mm) of pipe diameter. The trench width should be adequate to allow comfortable operation of pipe-cutting devices and to make reconnections. Extending the length of an excavation reduces the difficulties of insertion, especially in cold weather when the pipe becomes stiffer and more difficult to bend. The second excavation (the pull hole) should be large enough to accommodate pipe work, pulling cable angles, and possibly fusion equipment. Lining of large-diameter or deeply buried pipe may require a slightly larger excavation. Excavations should be sheeted and shored as required and bridged with steel plates during nonwork hours for public safety.

The distance between excavations varies depending on pipe layout and diameters. Single pulls in excess of 1,000 linear ft (300 m) are possible. It is recommended that excavations for service and lateral reconnections be made before beginning the pulling procedure.

8. The section of pipeline to be lined must be isolated by closing valves and then dewatered. (Water may actually make slip-lining easier, because HDPE pipe has nearly neutral buoyancy in water, but its presence may

cause difficulty for workers.) A section of existing pipe is then cut out of each end of the section to be lined, leaving a gap long enough to accommodate the pulling process at one end and the liner insertion at the other. Preparation of the carrier pipe may require pipe cleaning, depending on the extent of internal corrosion and tuberculation buildup. Also, sections of the old pipe must be cut out to facilitate reconnection of services and other laterals.

9. Once the carrier pipe is properly prepared, a winch cable is fed through it and attached to a bullet-shaped pulling head on the HDPE liner. The liner is constructed aboveground at a convenient location by joining sections of HDPE pipe with a butt-fusion machine. This machine produces a true joint with the same structural integrity and tensile strength as the pipe itself. Small-diameter liner pipe (up to 3.5 in. [89 mm] outside diameter) can be purchased in coils to reduce the number of field-fused joints; however, the added friction caused by pulling pipe that has attained a set curve during coiling must be considered. The liner is then pulled through the carrier pipe with a power winch. When installing large-diameter liners, a backhoe or bulldozer is often used in a push–pull technique, in which equipment pushing the liner assists the pulling action of the winch. Caution must be exercised during insertion to prevent gouging the exterior of the liner as it is pulled into the existing pipe.

10. After insertion is completed, the liner must be connected to the remaining water mains. Increasers can be fused to the end of the new liner and joined to old cast-iron pipe with approved mechanical fittings. Flanged plastic fittings, known as *stub ends*, with metal backup rings are available to bolt the new lining to existing or newly added steel pipe flanges.

    Service lines, fire hydrants, and branch laterals that have been excavated and cut out of the carrier pipe must be connected to the liner. Service corporations are thermally saddle fused directly to the liner and tapped with a self-tapping tee. Workers may have to cut in tees and make connections to existing branches for fire hydrant or sideline laterals, similar to those previously described for main line reconnections. Strict attention must be paid to the pressure ratings of all fittings.

11. The inserted and reconnected liner should be tested just as any newly constructed water main would be. Test pressure for a 3-hr test should not exceed 150 percent of the pressure rating of the pipe. HDPE pipe will expand in diameter during this type of testing, requiring regular additions of water to maintain the test pressure. Consult the pipe manufacturer for testing guidelines.

    Special attention must be given to exposed sections of the liner at pull or insertion excavations and at lateral reconnection points. Proper bedding and backfilling are necessary to prevent differential settlement.

12. Final completion of the lining process consists of restoring the main to service following disinfection, removing temporary bypass service, permanent paving, and site cleanup.

Drawbacks to the slip-lining method include a reduced cross-sectional area and the numerous excavations that must be made if many service and branch reconnections are involved. Regardless of these drawbacks, slip-lining is a useful lining method for the water utility engineer or manager.

MODIFIED SLIP-LINING TECHNIQUES

Scope

Modified or close-fit slip-lining techniques involve inserting a thermoplastic tube into the host pipe, which has been temporarily deformed to allow sufficient clearance for insertion. When the tube is subsequently returned to its approximate original shape and diameter, it provides a close-fit lining in the host pipe.

Two key differences distinguish this class of techniques from conventional slip-lining:

1. *Greater retention of hydraulic cross section.* Thin liners may actually improve the flow capacity of the existing pipe, despite the reduction in cross-sectional area, through improved smoothness (C value) and lack of joints in the liner.

2. *Flexibility of liner thickness.* Liner thickness can be selected to provide either fully structural (Class IV) or semistructural (Class II/III) internal pressure capability. The latter option can provide a more cost-effective system of renovation than other options for some types of host pipe deterioration.

Classification of Systems

Modified slip-lining systems can be classified into two main groups:

1. *Symmetrical reduction systems.* These systems involve use of a round thermoplastic tube with an outside diameter the same as or slightly larger than the inside diameter of the host pipe. The tube is passed through either a static die or an array of compression rollers, which temporarily reduce its diameter to allow sufficient clearance for insertion into the host pipe. After winch insertion, the tube is allowed to revert toward its original dimensions; in some cases, this reversion is accelerated by the application of internal pressure. These techniques generally use polyethylene pipe to take advantage of the material's molecular "memory" for the dimensions formed at the time of extrusion.

2. *Folded and formed systems.* These lining systems involve deformation of a thermoplastic tube into a folded C- or U-shape, either at the manufacturing plant or in the field (Figure 3-8). After winch insertion into the host pipe, the tube is returned to its original shape and diameter using heat and/or pressure. These techniques can be applied to polyethylene, fiber reinforced polyethylene, and polyvinyl chloride liner tubes.

Symmetrical Reduction Systems

A number of commercially available systems differ in the methods used to achieve the reduction in diameter, the methods and time scales for reversion to original shapes, and the role played by winch tension in the process.

Static die systems. Two such systems were originally developed in the United Kingdom by major gas and water utilities in response to the need for a trenchless renovation technique for municipal gas and water mains. Both involve pulling a length of polyethylene pipe through a static-diameter reduction die directly into the pipe to be renovated and maintaining winch tension until the pipe is fully inserted. When the winch tension is released, the polyethylene pipe rapidly reverts to its original diameter, simultaneously shrinking in length, until it achieves a close fit

Figure 3-8   HDPE pipe being formed into a C-shape on site

in the host pipe. The original process used an oven to heat the pipe before it passed through the die in order to reduce the winch forces required. Currently, this process is rarely used.

These processes have been mainly applied to cast-iron, ductile-iron, and steel gas and water mains in the diameter range 4 in. to 18 in. (102 mm to 457 mm). One process has been used, however, in water transmission pipe up to 42 in. (1,067 mm) in diameter.

Roller-based systems.   Two roller-based processes were developed to meet the needs of the US oil, gas, and mining industries for a thin polyethylene lining system to control internal corrosion in steel pipes carrying oil, gas, and aqueous products at high pressures. Both procedures involve pulling a length of polyethylene pipe through a series of reduction rollers, which may be hydraulically driven and/or braked, directly into the pipe to be renovated. The processes involve somewhat lower winch tensions than static die techniques require, and the liner may not revert toward original dimensions quite as rapidly when winch tension is released.

These processes have been used for insertion of long (1,000 ft [305 m]) sections of pipe in diameters up to 42 in. (1,067 mm). Most field experience has involved lining welded steel pipelines, although some installations have inserted linings in cast-iron municipal gas and water lines.

A third roller-based process, developed in the United Kingdom in parallel with the die-based processes, has been extensively used for renovation of cast-iron gas mains, and subsequently water mains, in the diameter range of 4 in. to 18 in. (102 mm to 457 mm). The process involves pushing a length of polyethylene pipe through a series of reduction rollers. In contrast to the static die systems and the other roller-based systems, a large part of the reduction in diameter is retained for a period ranging from hours to days, depending on ambient temperature. Full reversion to final dimensions requires the application of a high internal water pressure for 12 to 24 hr. This time lag allows insertion of the reduced pipe into the host either directly, as with the other processes, or at a different time and location.

General characteristics. All of the symmetrical reduction techniques share the following characteristics:

1. The energy required to reduce the polyethylene pipe diameter increases dramatically with pipe size and wall thickness (indicated by SDR). Therefore, winch tensions and/or hydraulic power requirements may rise with increasing diameter and thickness, limiting the maximum pipe wall thickness that can be handled at each diameter. These limitations vary from process to process and affect the ability of the processes to be used for renovating thick, Class IV polyethylene liners.

2. Care is needed in lining pipes with significant local variations in internal diameter because of manufacturing tolerances or other potential obstructions such as joint offsets. These variations may reduce or even eliminate insertion clearance at certain points.

3. The required polyethylene pipe diameters are rarely standard pipe sizes. Manufacturing pipes in needed sizes may involve the use of special extrusion dies, with potential implications for the cost and timing of projects.

4. The processes are unable to negotiate manufactured bends, and local excavation is required at the locations of these service connections and any other in-line fittings.

5. Sufficient site space is required to accommodate butt-fusion welding of polyethylene pipes into long sections prior to diameter reduction and during subsequent insertion.

Folded and Formed Systems

Factory folded/hot re-rounded systems. These systems involve tubing heated and folded into a C- or U-shape in the factory and then transported to the job site on a reel that may contain up to 2,000 ft (610 m) of liner, depending on diameter. The folded liner is winched into the host pipe (in some cases, after reheating the liner on the reel to facilitate placement) (Figure 3-9). Then it is re-rounded using a combination of heat (usually provided by steam) and pressure (provided by steam and/or air). The re-rounding process may be progressive, using a device that is propelled through the liner, or it may occur simultaneously throughout the pipe length.

Two such systems use polyethylene pipe to line mains ranging from 4 in. to 16 in. (102 mm to 406 mm) in diameter with liner pipe thickness in the range SDR 21.0 to 32.5. In a recent development, large-diameter (21 in. [533 mm]) folded polyethylene pipes have been shipped to job sites in 40-ft (12-m) lengths, where factory-matched lengths are joined by butt-fusion welding into long (1,500 ft [457 m]) sections prior to insertion and re-rounding. This technique allows use of larger diameters than are possible with the more conventional reels of folded material. Such systems have been widely used for wastewater pipes, and they have been evaluated for gas and potable water lines.

In addition, at least two slip-lining systems use modified rigid polyvinyl chloride, mainly for wastewater pipelines. The manufacturers of these systems are currently exploring their potential for use in water pipelines.

A more recent addition to the available product range is a polyethylene liner reinforced with a circular woven polyester yarn (PRP). This system has been developed in the United Kingdom specifically for renovation of water pipes, and it is

Figure 3-9   Folded and banded HDPE pipe is winched into the host pipe

currently available in diameters ranging from 3 in. to 6 in. (76 mm to 152 mm) with probable extension to 8 in. (203 mm). Reel-wound PRP liner is delivered on site in folded form and then winched at ambient temperature into the host pipe. It is then rapidly re-rounded using a combination of air and steam. The process leaves a structural Class IV liner in place with a long-term independent pressure capability of 150 psi (1,034 kPa) and a wall thickness of only 0.12 in. (3 mm). The liner can negotiate manufactured bends of up to 45° with some internal wrinkling.

Site cold-folded/cold re-rounded systems.   These systems involve passing lengths of thin, round polyethylene pipe through a piece of site-based equipment that folds the pipe into a U-shape. The shape is restrained by a series of thin plastic straps applied to the folded pipe as it exits from the folding machine. After winching into the host pipe, the polyethylene liner pipe is re-rounded through the application of internal pressure, which breaks the straps.

This method is generally applicable to pipe thicknesses up to SDR 26. It has been used in Europe as a Class III liner for water pipes up to 36 in. (914 mm) in diameter. It offers the advantage over factory-folded techniques that additional lengths of polyethylene pipe can be fused to the lining section at any time prior to passage through the folding equipment and during insertion of the liner into the host pipe. Hence, the method can accommodate long insertions at large diameters with minimal site space problems. The polyethylene pipe is normally of nonstandard size produced with outside diameter equal to the minimum expected inside diameter of the host pipe.

General characteristics.   All of the folded and formed techniques share the following characteristics:

1. They involve much larger insertion clearances and are not as time-sensitive as symmetrical reduction techniques, hence they are much less sensitive to local variations in pipe diameter.

2. The folded shape generally allows insertion through a smaller entry pit than the symmetrical reduction techniques require.

36 REHABILITATION OF WATER MAINS

3. Where such systems are installed as Class IV liners, confirm that the folding and re-rounding process has not affected the long-term pressure capability of the liner.

CURED-IN-PLACE LINING TECHNIQUES

Scope

Cured-in-place pipe (CIPP) lining techniques involve inserting a polymer fiber tube or hose impregnated or coated with a thermoset resin system into the host pipe. The resin is then cured, either under ambient conditions or by application of heat using steam or water, to produce either a rigid "pipe within a pipe" or a semi-rigid liner which depends on adherence to the pipe wall for support.

Techniques of this type have been extensively used for rehabilitation of municipal and industrial wastewater pipelines and municipal gas lines. Applications to potable water lines have been limited by the need to obtain appropriate approval from relevant health authorities. However, variants of some of these systems have now been approved by NSF for use in North America.

Classification of Systems

CIPP systems can be classified into three main groups:

1. *Felt-based systems.* The lining tube is produced from a nonwoven polyester felt coated on one face with a layer of elastomer. Varying the thickness of the felt tube and/or including reinforcing fibers allows the lining, when impregnated with resin and cured, to meet a wide range of design requirements.

2. *Woven hose systems.* The lining tube consists of a circular woven, seamless, polyester fiber hose coated on one face with a layer of elastomer. The construction of the hose is tailored to meet internal pressure requirements, and the cured resin layer serves merely as an adhesive to the host pipe.

3. *Membrane systems.* The lining tube consists of a very thin elastomeric membrane designed only to offer internal corrosion protection and bridge very small pinholes and joint gaps. The cured resin layer again serves only as an adhesive to the host pipe. One variant joins such a membrane with a woven hose for a combination that can span larger holes and gaps.

Systems are also available based on combinations of these types.

Felt-Based Systems

The original felt-based lining system, developed in the United Kingdom in 1971, has since been used to rehabilitate many thousands of miles of municipal and industrial wastewater pipes throughout the world. The system has been utilized in pipe ranging from 4 in. to 108 in. (102 mm to 2,743 mm) in diameter, and it can accommodate noncircular shapes and negotiate 90° bends. A number of similar processes have evolved over the years.

Felt-based systems can be classified according to

1. *Installation method.* Some systems, like the original process, are installed by inversion, in which the impregnated tube is simultaneously fed through the pipe and turned inside out by water pressure. Some are installed by pulling lining tubes into pipes and inflating them (Figures 3-10 and 3-11).

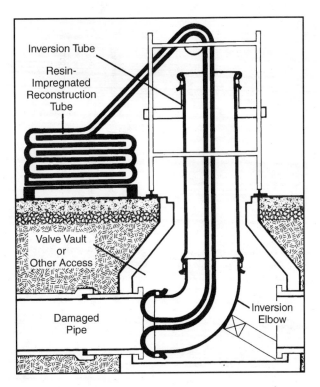

Figure 3-10 Felt tube fed into the pipe

2. *Tube material.* Systems vary based on whether the felt tube also incorporates textile or glass reinforcing fibers.

The tube is manufactured to suit specific host pipe dimensions and impregnated with the appropriate resin, either in the contractor's factory or, for large-diameter tubes, on site. The resin used—either polyester, vinyl ester, or epoxy—is selected to meet specific chemical and mechanical performance requirements. For linings in potable water pipes, the requirements of NSF approval determine the type of resin and coating.

The impregnated tube is normally cooled in ice and transported to the job site in a refrigerated truck to prevent premature setting of the resin. The tube is then either inverted into the pipe using static water heads or an air-over-water pressure system. Alternatively, it may be winched into the pipe and then inflated by inverting through it a sizing tube using air or water pressure. The liner is then cured for 6 to 9 hr by either heating the inversion water or using steam (Figure 3-12). After curing, the ends of the liner are cut, and in the case of pressurized potable water lines, suitable end seals are fitted.

The thickness of a lining for wastewater pipes is determined mainly by the need to resist buckling due to external water head. In lining potable water pipes, resistance to internal pressure is the key design parameter. Currently available systems offer Class III semistructural capability at normal water system operating pressures. The systems can also be designed to offer significant inherent resistance to external head, so their use may be indicated in conditions with known risk of external water head, and where subsequent service outages, line depressurization, or transient vacuum conditions are likely to occur.

38   REHABILITATION OF WATER MAINS

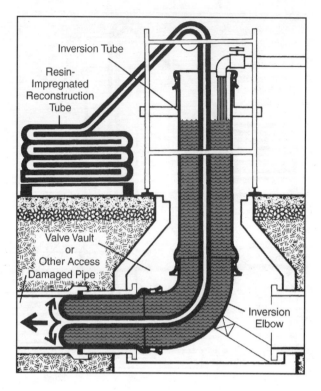

Figure 3-11   Filling the tube-feeding standpipe with cold water

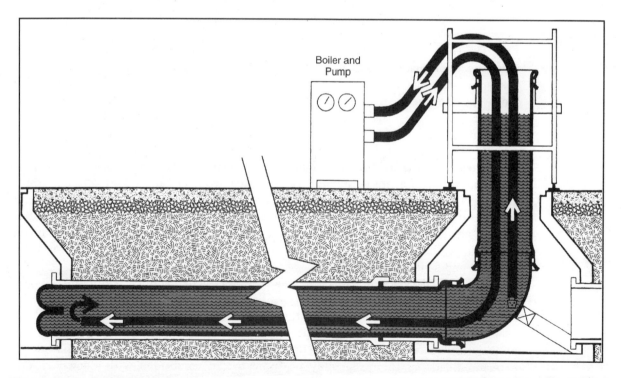

Figure 3-12   Heating water after the tube is in place, causing the resin to adhere to the pipe walls

Woven Hose Systems

These systems utilize circular woven polyester fiber hose coated on one face with a layer of elastomer. Prior to installation, the uncoated (inner) face of the hose is impregnated with a thin layer of epoxy resin and then inserted in the pipe by an inversion technique using either air or water pressure. During insertion, the liner is turned inside out so that the resin is pressed against the inner surface of the host pipe. After heat curing (typically by circulating hot water or steam for 8 hr), the resin serves to adhere the liner to the pipe wall. The system is capable of negotiating a sequence of 90° manufactured bends. The range of available diameters is typically 4 in. to 40 in. (102 mm to 1,016 mm).

Woven hose systems generally offer Class II semistructural capability at typical operating pressures, so their use is indicated in pipes suffering from severe internal corrosion and pinhole leakage, leakage due to faulty joints, and in some cases, problems arising from localized external corrosion. The installed liner is very thin, and its high C value and joint-free construction may allow flow rates in the rehabilitated pipeline identical to those in the original pipe in new condition.

Resistance to external buckling loads primarily depend on the quality of the adhesive bond to the pipe wall, and hence on the quality of cleaning achieved prior to insertion. However, variants of the woven hose system are at least self-supporting (i.e., Class III) when cured, reducing their dependence on the condition of the internal pipe surface.

Woven hose systems were originally developed in Japan for rehabilitation and earthquake protection of municipal gas lines. The systems have also been used in Japan and Europe for rehabilitation of potable water pipes.

Membrane Systems

These systems line pipes with thin elastomeric membranes coated with thermoset resin. Liners are inverted into the host pipe using air pressure and heat cured in a manner similar to woven hose systems. Membrane lining systems were developed for rehabilitation of leaking low-pressure (up to 10 psi [69 kPa]) gas mains. They offer more limited capabilities for spanning holes and gaps at typical water system operating pressures than woven hose systems.

Bibliography

Accelerated Age Testing of Epoxy Resin. 1990. Medmenham, U.K.: Water Research Centre.

Deb, Arun K., Yakir J. Hasit, and Chris Norris. 1999. *Demonstration of Innovative Water Main Renewal Techniques.* Denver, Colo.: AWWA Research Foundation and American Water Works Association.

In-situ Epoxy Resin Lining—Operational Guidelines and Code of Practice, 2nd ed. 1990. Medmenham, U.K.: Water Research Centre.

This page intentionally blank.

AWWA MANUAL M28

Chapter 4

Internal Joint Seals

Internal joint seals are designed to seal the inside surfaces of leaking pipe joints in all types of 16-in. (406-mm) diameter or larger pipes. Internal joint seals eliminate leaks in pipelines with internal working pressures up to 300 psi (2,068 kPa). (Special seals can be engineered to accommodate higher pressures.)

An internal joint seal incorporates a number of lip seals on its outer edges that together completely seal the circumference of the pipe on either side of the joint. The seal's flexibility ensures a bottle-tight seal around the entire pipe joint, while its low profile and graded edge allows water to flow without creating turbulence.

All materials used in internal joint sealing of potable water mains should be classified by NSF International (NSF) or Underwriters Laboratories Inc. in accordance with Standard ANSI/NSF 61.

Internal joint seals are manufactured to specifications from an ethylene propylene diene monomer (EPDM) synthetic rubber compound. They are designed to seal internal surfaces of joints in all types of potable water and source water mains, from 16 in. (406 mm) in diameter upward with operating pressures up to 300 psi (2,068 kPa). See Table 4-1.

Seals are packed for shipping in a manner that will not damage or deform them and special precautions must be maintained until the seal is fitted within the main. Seals must be stored in a dry environment at room temperature, and they should not be allowed to remain in direct sunlight.

Prior to fitting, seals should undergo a thorough visual examination by operators, paying particular attention to the ribbed (lip seals) sections of the seal. If in doubt, do not use a seal.

Internal joint seals are available in two widths: standard to accommodate joint gaps up to 5 in. (127 mm), and extra wide for joint gaps up to 8 in. (203 mm). (Special seals can be engineered to accommodate a variety of joint gap widths.)

FITTING PROCEDURE FOR INTERNAL JOINT SEALS INSTALLED IN WATER MAINS

All procedures covering safety of personnel working within a pipeline must normally be carried out in accordance with applicable safety regulations.

42 REHABILITATION OF WATER MAINS

Table 4-1   Material details for internal joint seals*

A.	Retainer bands: Type 304 stainless steel, $^3/_{16}$ in. (4.8 mm) TK × 2 in. (51 mm) wide strips; ultimate tensile (85,000) yield strength (35,000) meets ASTM A-167, ASTM A-267, ASTM A-479-85
B.	Shims: Type 304 stainless steel, 0.048 in. (1.22 mm) TK × 2 in. (51 mm) wide × 6 in. (152 mm) long
C.	Test valve, low profile: Type 304 stainless steel
D.	Plastic backing band: Material meets ASTM specification D-1248 Sub Type 3. This material is a high-density polyethylene extrusion with a specific gravity of 0.960.
E.	Nontoxic thread seal compound: La-Co Slic Tite Paste w/Teflon or equivalent
F.	Nontoxic vegetable lubricant: Tyton Joint or equivalent
G.	Joint gap filler: Portland cement mortar, Type 5 meets ASTM C-150 epoxy for underwater repair and grouting
H.	Pipe surface gel: Aquatapoxy nontoxic patching gel or equivalent
I.	Internal joint seal: Certified to ANSI/NSF Standard 61 EPDM rubber for potable water

*Internal joint seals meet American Society for Testing and Materials (ASTM) specification D-2000.

1. *Pipeline preparation.* Experience has shown that most deleterious deposits can be removed from pipe walls around a joint by hand scraping and brushing. Occasionally, power tools are used to remove stubborn deposits or hard-scale lamination. Whichever method is adopted, the main must be as clean as reasonably possible to provide an acceptable working environment for the operators.

2. *Joint filling.* During pipe cleaning operations, the gap between the joints must be cleared of dust and debris leaving a clean area for joint filling. The joints are filled with portland cement to the full depth of the gap and rendered flush with the internal surface of the pipe. All surplus material and spillage should be removed from the joint area prior to the surface preparation for the seal. The joint filling operation should always be carried out before final preparation.

3. *Surface preparation of the joint area.* The area of pipe on either side of the joint where the actual lip seals make contact must be prepared. The resulting finish must allow the lip seals to bed consistently so they provide a permanent seal (Figure 4-1).

   a. All high and low spots that form surface imperfections running axially through or part way through the sealing surface must be removed. Deep imperfections must be filled with approved compounds. The material must be smoothed to match the surface of the joint area.

   b. Circumferential grind marks are allowed, provided they do not exceed 0.030 in. (0.76 mm) in depth.

   c. The extent of the prepared area on either side of the joint must be compatible with the lip seals, and the prepared area must extend at least 1 in. (25 mm) beyond either side of the ribbed section of the seal.

   d. The pipe should be marked with grease chalk to define the prepared areas and seal position.

   Note: The importance of good surface preparation cannot be overemphasized.

INTERNAL JOINT SEALS 43

Figure 4-1 Joint area is cleaned and prepared prior to installation of the seal

Figure 4-2 A nontoxic lubricating soap is applied and the seal is carefully positioned with its retaining band

4. *Surface lubrication.* **Immediately** prior to fitting the seal, the area must be cleaned with a dry brush and coated with a nontoxic lubricating soap compatible with the composition of the internal joint seal. The lubricant is hand applied over the entire prepared area using an ordinary paintbrush. Care must be taken not to pick up dust deposits from the unprepared surface and deposit them into the lubricant. The lubricant is purely an aid to fitting the seal and in no way contributes to its sealing capabilities. Before using a lubricant product, confirm its acceptability for use in contact with potable water (Figure 4-2).

5. *Positioning the seal.* Confirm that the seal is undamaged and the test unit (stem and backnut) is tightened to a torque of 15 to 17 in.-lb (1.7 to 1.9 N·m). The internal joint seal is placed in position bridging the joint gap, guided by the previously drawn chalk marks. The seal must be positioned accurately on the prepared areas. The test unit in the seal must be located at either the 9 or 3 o'clock position.

The test unit mentioned is generally referred to as a *valve*. This assembly has no integral sealing capability. It is sealed only by the fitting of the plug. Position the backing band behind the seal over the joint area.

44   REHABILITATION OF WATER MAINS

This component is sometimes easier to install after one of the retaining bands has been set in place.

6. *Positioning retaining bands.* Before stainless steel retaining bands are placed in the grooves provided in the seal, two 0.048 in. (1.22 mm) stainless steel radial shims are placed in these grooves at the band gap to provide a bridge that will continue the radial load transmitted to the internal joint seal when the bands are expanded. Both bands are temporarily locked in position, with their ends equally spaced across the spring steel shims. Band dimensions and material vary depending on diameter and pipeline product, but a common unit is 0.1875 × 2 in. (4.8 × 51 mm) 304 or 316 stainless steel.

7. *Expanding the seal into position.* A hydraulic expanding device called a *ring expander* is used to apply the correct pressure to the retaining bands.

   a. When positioning the expander in line with a retaining band, ensure that the band remains in the groove of the seal and does not become moved or dislodged. Also ensure that the expander is positioned correctly in both planes. For example, should the expander be placed by touching the bands at the invert, it is possible to lock the expander at full pressure without exerting any load at all at the top of the seal.

   b. The expander is expanded radially with a predetermined pressure (not more than 4,500 psi [31,027 kPa] on the pump gauge) transmitting the required load against the retaining band and the internal joint seal. Table 4-2 gives band pressures for various seal sizes. This pressure is held for at least 2 min.

   c. A space provided in the expander exposes the cleats of the retaining band. A locking piece called a *wedge* is fitted between the exposed gap of the expanded band ends. The size of the wedge is selected that gives a slight interference fit between the band ends. The wedge is tapped leading edge first into position, locking in the compression of the internal joint seal. The radius of the wedge is matched to suit the pipe diameter.

Table 4-2   Retainer band expansion pressures

Diameter	Pneumatic Expander	Hydraulic Expander
16 in. (406 mm)	400 psi (2,758 kPa)	2,700 psi (18,616 kPa)
18 in. (457 mm)	400 psi (2,758 kPa)	2,700 psi (18,616 kPa)
20 in. (508 mm)	400 psi (2,758 kPa)	2,700 psi (18,616 kPa)
24 in. (610 mm)	400 psi (2,758 kPa)	2,700 psi (18,616 kPa)
30 in. (762 mm)	400 psi (2,758 kPa)	3,800 psi (26,200 kPa)
36 in. (914 mm)	400 psi (2,758 kPa)	3,800 psi (26,200 kPa)
42 in. (1,067 mm)	400 psi (2,758 kPa)	3,800 psi (26,200 kPa)
48 in. (1,219 mm)	400 psi (2,758 kPa)	4,000 psi (27,579 kPa)

Notes: 16 in. (406 mm) through 48 in. (1,219 mm) bands are one-piece units.

Caution should be used during hydraulic expanding of retainer bands; hydraulic expansion of the retainer bands requires a slower rate of expansion due to cold flow of seal.

Apply expansion pressure at a slow rate until maximum allowed gauge pressure is reached. Rapid expansion will result in damage to retaining bands.

54 in. (1,372 mm) through 144 in. (3,658 mm) retaining bands expand at 4,000 psi (27,579 kPa) for two- and three-piece bands. Apply pressure at all wedge points.

INTERNAL JOINT SEALS   45

d. The pressure is released from the expander and the procedure is repeated on the second retaining band of the seal.

e. This entire re-expansion operation must be completed after at least 1 hr has elapsed after the first expansion. This precaution allows for any seal relaxation that may take place, in which case a slightly wider wedge can usually be fitted. The load forces transmitted by the expander have been determined from test data and should not be altered by changing the pressure over the center of the seal (Figure 4-3).

8. *Extra-wide seals*. Fitting procedures are identical to those of standard seals. Sometimes an extra retaining band is placed over the center of the seal.

9. *Testing the seal—Test 1*. Two individual pressure tests are applied to each seal. In the first test, air is introduced at a pressure of 5 psi (34 kPa) into the seal through the valve in the internal joint seal. This pressure is sustained while soapy water is applied to the outer edge and entire body of the seal. Any leak, indicated by a growing bubble or stream of bubbles, must be stopped.

10. *Testing the seal—Test 2*. The second test is applied after each section has been completed and the seal has had time to set. Air is introduced at a pressure of 10 psi (69 kPa), but a restraining device is fitted over the center of the seal to prevent excessive ballooning to the membrane that will occur at this higher pressure. Soapy water is again used to detect any leaks, which must be sealed.

11. *Testing an extra-wide seal*. Testing procedures similar to those for standard seals should be carried out. Care must be exercised, however, to ensure that the test pressure does not exceed 5 psi (34 kPa), if the test is conducted without the central restraining device in order to check the body of the seal. At higher pressures, excessive ballooning will occur, resulting in movement of the sealing faces. On seals with very large diameters and where multisection retaining bands have been used, it is not possible, or desirable, to conduct tests above 5 psi (34 kPa) because no central restraining device is available (Figure 4-4).

12. *Completed seal and completion report*. The completed seal provides a permanent solution to leaking joint problems. Finally, a completion report should be supplied detailing the type and location of each seal and any other repair/rehabilitation activities.

46 REHABILITATION OF WATER MAINS

Figure 4-3   An expansion ring is placed over each retaining band and a wedge is inserted between the band ends

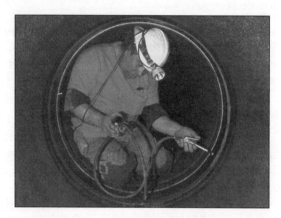

Figure 4-4   Each seal is leak tested twice before the main is put back into service

AWWA MANUAL M28

Chapter 5

Pipe Bursting

Pipe bursting is a trenchless method of replacing existing water mains by breaking and displacing existing pipe and installing a replacement pipe in the void created. The pipe-bursting process replaces the original pipe with a new pipe that is the same diameter or larger.

The system consists of a pneumatic, hydraulic, or static bursting unit that splits the existing pipe while simultaneously installing a replacement pipe of the same or larger diameter and pressure rating. The pipe-bursting tool is designed to force its way through existing pipe by fragmenting or splitting the pipe and compressing the materials into the surrounding soil as it progresses.

Pneumatically or hydraulically powered equipment has either front or rear expanders and a sleeve for the connection to the replacement pipe. It is used in conjunction with a 5-, 10-, or 20-ton constant-tension variable-speed winch. The size of the winch depends on the diameter of the pipe to be replaced. The pulling force should be maintained at a value less than the tensile strength of the replacement pipe to avoid overstressing the new material.

A static bursting unit is either pushed or pulled with solid steel rods or pulled by heavy chain or cable. Care should be exercised not to overstress the new pipe with this method.

The bursting action of the tool increases the external dimensions of existing pipe sufficiently to break it into pieces, which the tool compresses into the surrounding ground. This action not only breaks the pipe but also creates the void into which the bursting unit is pulled or pushed, allowing forward progress. At the same time, the replacement pipe moves forward by direct attachment to the sleeve on the rear of the bursting unit. As the pipe-bursting tool advances through the existing pipe, the replacement pipe advances directly behind to fill the void created.

A pneumatic bursting unit supplies its own forward momentum with assistance from a winch, which guides the bursting unit. A hydraulic bursting unit expands and contracts while being winched forward, creating a void for the new pipe (Figure 5-1).

These systems work with existing pipe varying in diameter from 4 in. (102 mm) to 48 in. (1,219 mm). Pipe to be replaced can be either fracturable material or material that can be sliced by cutters integrated into the bursting unit.

48 REHABILITATION OF WATER MAINS

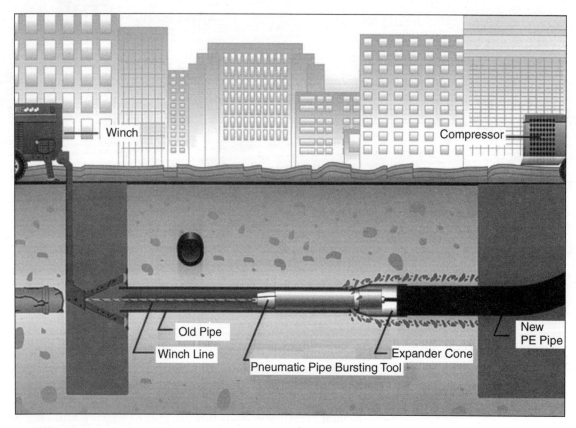

Source: Courtesy of TT Technologies Inc.

Figure 5-1   Diagram of a typical pneumatic pipe-bursting operation

A review of plans should identify service connections, valves, hydrants, and fittings, so they can be located, excavated, and exposed before pipe-bursting operations commence. In addition, closed-circuit television can be utilized to accurately locate fittings, valves, repairs, etc. A temporary bypass system may be needed to maintain service to consumers.

All service connections should be completely disconnected and isolated from the existing pipe before pipe-bursting operations commence. Service connections should not be reconnected to the replacement pipe until installation, disinfection, and testing are complete.

All pipes and underground structures crossing within 3 ft (1 m) of the pipe to be burst should be excavated and exposed. The soil between the pipe to be burst and other structures must be removed so bursting forces are not transmitted through the soil to the adjacent structures.

Cleaning of existing pipe is not necessary for most pipe-bursting processes. Sediments in the existing pipe are simply pushed into the surrounding soils.

Pit excavation is needed to accommodate replacement pipe sections. The operation must provide for removal and disposal from the project site of all valves, fittings, paving, trees, vegetation, and other known obstructions. Pits should be excavated at the ends of the lines to be replaced unless other locations are required. Pits should be centered over the existing line, and excavation sizes should be verified in the field prior to construction of the project. They may be subject to review.

Polyethylene pipe is a common choice for replacement pipe. Sections should be assembled and joined on the job site by the butt-fusion method. This process should be carried out in the field by qualified operators with prior experience in using proper jigs and tools according to procedures outlined by the pipe manufacturer and the equipment manufacturer. Every joint should have a smooth, uniform, double-rolled back bead made while applying the proper melt, pressure, and alignment. Care must be taken not to damage the inserted pipe as it passes through the fragments of the old pipe. The contractor should accept sole responsibility for providing an acceptable butt-fusion joint. All joints should be made available for inspection by utility staff before the pipe bursting operation begins.

Testing and disinfection should follow installation of the replacement pipe, according to the same procedures used for new pipe installation. After the replacement pipe has been completely installed, disinfected, and tested, all existing active services indicated on the contract plans or identified by the utility should be reconnected to the replacement pipe.

The contractor who provides the pipe-bursting services should be certified by the equipment manufacturer. Contractors who bid on this type of project should be selected from a list of prequalified contractors, if allowed by law. The qualifications submitted by a pipe-bursting contractor should include detailed descriptions such as

- Name, business address, and telephone number, as well as the names of all supervisory personnel to be directly involved with the project

- List of similar projects completed, including name, address, and telephone number of the project manager or other contact, as well as descriptions of the types of pipes burst and the sizes and lengths of replacement pipes installed

The contractor should sign and date the information provided and certify that the information is true and accurate and that the supervisory personnel identified will participate directly in the pipe-bursting project. Substitutions of personnel or methods should not be allowed without written authorization of the utility.

The contractor should submit written descriptions of the construction methods and equipment to be used, as well as pit dimensions and locations required for equipment and material access.

Pipe-bursting technologies are subject to patent protection, so the contractor should warrant to the utility that the equipment to be used is furnished in accordance with applicable licensing or use agreements and that the prices quoted cover all applicable royalties and fees required under such agreements. The contractor should protect the utility against any costs, loss, damage, or expense arising out of any claim of infringement of patent or trademark or any violation of a licensing agreement.

This page intentionally blank.

AWWA MANUAL M28

Chapter 6

Maintaining Service

The purpose of any water supply system is to maintain a continuous supply of safe water sufficient for customers' needs, including fire protection. This charge remains of primary importance even as rehabilitation work is performed. As noted earlier in this manual, depending on the cleaning or lining method used, temporary distribution system shutdowns may occur. Some processes require relatively brief shutdowns, so work can be completed without installing bypass lines. Conversely, some cleaning techniques and all lining methods require more extensive shutdowns that may create the need to install bypass piping.

BYPASS PIPING

Installing bypass piping must be a carefully planned and well-coordinated procedure, and each project has unique considerations. The utility and contractor should jointly review the plan, keeping in mind customer service issues and traffic concerns. Other considerations include 1) individual service connections, 2) overall demand, and 3) fire protection demand (e.g., pipe sizing). Note that bypass lines are connected to fire hydrants or other temporary connections outside the area of the shutdown.

The installation of bypass lines can be the most time-consuming and labor-intensive operation of a cleaning or lining project. However, the use of bypass piping does allow fairly long shutdowns while still maintaining acceptable service to consumers. The hydraulic requirements of the portion of the water system to be removed from service establish the parameters for sizing the bypass piping network. Temporary connections may be required for fire protection, depending on local jurisdiction. After the piping has been sized and installed, it must be disinfected as directed in AWWA C651 before being placed in service. Connections should always maintain positive pressure in the bypass pipes. Leakage should be minimized to limit damage, save money, and maintain customer confidence.

Residential Installations

A residential area bypass line is usually 2 in. (50 mm) in diameter with provisions for a 0.75-in. (19-mm) or 1-in. (25-mm) hose connection to each residence along the main (Figure 6-1). The bypass hose is connected to the customer's service line through a

51

52 REHABILITATION OF WATER MAINS

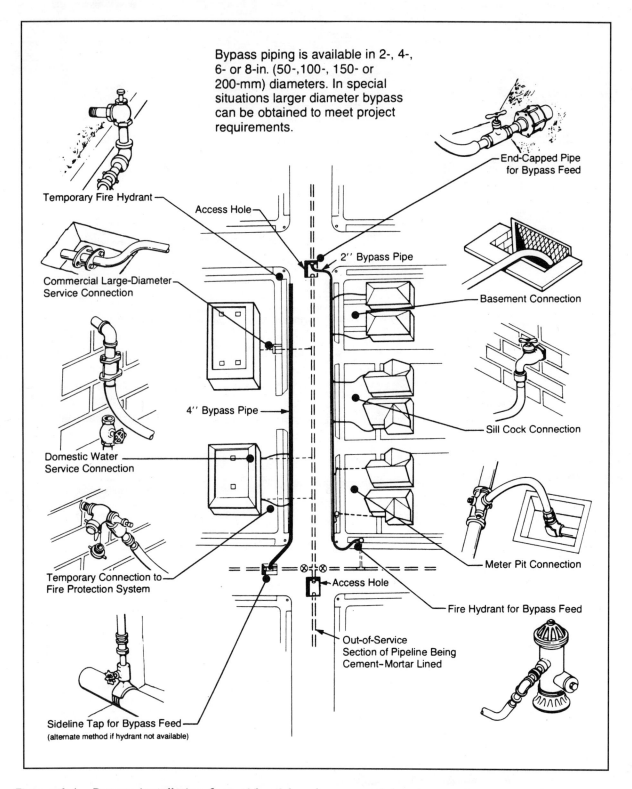

Figure 6-1 Bypass installation for residential and commercial water service

meter pit, an outside hose bib on the dwelling, or interior pipes accessed through a basement window. The bypass line is usually installed along the curb or in the gutter so that it can be buried under driveway aprons and street intersections. Cold-mix asphalt can be used to cover and protect bypass piping in traffic areas. The exact location of the bypass line installation depends on project-specific circumstances. Site conditions affect the amount of potential damage to the line, possible tripping hazards, and any obstructions to pedestrians and vehicular traffic.

Commercial Installations

In a business or commercial area, bypass lines are usually 4 in. (100 mm) or larger in diameter. Additional precautions may be required to avoid problems with parallel parking of vehicles and damage to the bypass line or vehicle tires when parking or turning. Damage usually occurs because of in-line taps, couplings, or valves. Building connections are made through hoses similar to residential installations, except that larger hoses may be required. Depending on local street and traffic concerns, bypass piping may traverse driveways and street intersections by 1) burying temporary pipes in a pavement trench, 2) placement on the street surface with asphalt ramping, 3) boring under the street, or 4) running pipe through existing culverts.

Reconnection of Service Lines

Once bypass piping has been used to maintain service during cleaning or lining projects, care during the reconnection phase is very important to avoid further service interruptions. Upon completing the cleaning or lining, the permanent pipelines must be flushed and disinfected according to AWWA C651. Following this step, the lines are ready for service, the bypass lines can be disconnected, and customers reconnected to the permanent service lines.

Some cleaning and lining operations may require that individual service lines be blown back to the main before reconnection; this can be accomplished by various means. High-pressure water can be forced back through the service line from inside the building to the main via a hose-and-pump connection. Blowback can also use compressed air fed directly from the compressor or from portable, worker-borne tanks. Note that all blowback air should be generated by oil-free compressors.

Problems

Exposed bypass lines may be subject to overheating problems during the summer or year-round in more southerly areas of the country. In these situations, the bypass line should be coated with white or other heat-reflective paint to minimize solar heat gain. Conversely, during the winter and in colder areas of the country, freezing of temporary pipes may present problems. If scheduling permits, rehabilitation or cleaning projects should be conducted during more temperate months. However, if work during extreme cold weather is unavoidable, provisions must be made to maintain water flow in all bypass lines to prevent freezing.

Unfortunately, the use of bypass piping may be the aspect of a cleaning or lining operation most likely to generate vehicle damage or personal injury claims. The temporary network also represents the greatest potential source for emergency repairs due to hose ruptures or vandalism. As discussed in the next chapter, bypass service increases the need for good communications with various groups of consumers.

SUMMARY

Any of the methods outlined in this manual can be performed while safe, adequate water service is maintained to the customer base. Some methods may require more extensive temporary service installations than others, and the economics of the entire project may be greatly affected by the temporary service methods chosen.

AWWA MANUAL M28

Chapter 7

Customer Relations

Perhaps the most important operational or maintenance aspect of any water supply system is the relationship that develops between the utility and its customers, in part through efforts to maintain continuous water service. In this regard, both residential and commercial/industrial customers must understand the general nature of any projects that may affect their water supply. This information should be provided in a timely and effective manner prior to a scheduled project. Advance notice allows customers to ask questions and resolve problems prior to beginning the actual work. Planning and implementing a good customer relations program is equally as important as planning the actual cleaning or lining work.

CUSTOMER NOTIFICATION PROGRAM

When considering communication needs prior to a cleaning or lining project, early planning of the customer notification program is an important priority. The simple objective of this communications program is to develop customers' understanding of how the project will impact their water supply.

The first step in the program is to determine who will be affected by the cleaning or lining project. An overall project schedule should be determined so that advertising and public notices can appear before and during the actual work. This coordination requires careful attention to publishing schedules of daily and weekly newspapers as well as radio and TV station broadcast schedules. Contact and project information can also be included on a utility's Internet page, if available.

Any type of notification should stress the necessity of the project and explain what will happen. A letter describing the project, including details of detours and local area access, is often helpful (Figure 7-1). Letters should be delivered to all residential and commercial customers within and near the project area. Additionally, doorknob hanger cards that explain the project can be used.

Most utility work causes some traffic disruption and the need for temporary detours. Detours should be identified and public notifications prepared for newspaper publication so that the traveling public (and local rapid transit district) is aware of the work and can avoid the area. Also, the work area should be properly barricaded and caution notices posted (Figure 7-2).

56   REHABILITATION OF WATER MAINS

Hackensack Water Company
200 Old Hook Road
Harrington Park, N.J. 07640

Dear Customer:

Starting the week of July 16, Hackensack Water Company will be cleaning and lining a major water main which supplies portions of your community. Following the route indicated on the map below, construction will begin in Englewood at the intersection of Van Nostrand Avenue and Grand Avenue, and will proceed south on Grand Avenue through Leonia, Palisades Park, and Ridgefield, to end at the Hendricks Causeway. The project is expected to take three to four months to complete.

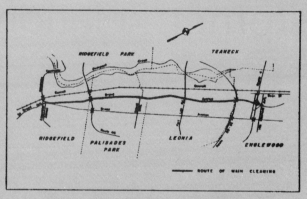

IMPROVING YOUR WATER SYSTEM

As part of our ongoing maintenance program, we must periodically recondition sections of our water system, which assures you, as our customer, continued dependable water service. Pipeline reconditioning involves scraping the inside of the main to remove accumulated materials, followed by the application of a thin coat of cement to protect the water main from further corrosion. This work results in improved water flows in the pipeline and more efficient operation of our system. As a customer, your benefit is the assurance that water pressure in your home or business, as well as essential fire protection, will be maintained at all times.

HOW WILL THIS WORK AFFECT YOU?

While the work is in progress, residents and businesses along the route will be provided with a temporary water supply since the water main must be taken out of service for cleaning. Connections will be made to your service line from alternate pipelines laid alongside the curbs on each side to the street. We will make every effort to ensure that your normal water service is maintained at all times.

PLEASE...WATCH YOUR STEP

If you live, travel or shop in the area, you'll notice a number of openings in the street to allow access to the water main and to accommodate the bypass pipelines running across the intersection and driveways. All openings will be covered during the work and will be resurfaced as soon as possible after the project is finished. Every possible safety precaution will be taken and all work areas will be barricaded and carefully marked to prevent injury. You can help by cautioning your family and friends to watch their step and avoid the work areas as much as possible.

INCONVENIENCE TEMPORARY...IMPROVEMENTS PERMANENT

We realize our main cleaning project may cause some inconveniences for those who live and work in the area. We are working closely with your local officials and police department to minimize traffic disruptions and we will do our best to see that the construction goes smoothly.

IF YOU HAVE A PROBLEM...

If a problem should occur as a result of our work, please report it directly to Hackensack Water Company by calling our 24-hour emergency switchboard at 487-0011.

Thank you for your patience and cooperation while we work to improve your water system.

Hackensack Water Company

Figure 7-1   A sample letter notifying consumers of work to be done

CAUTION

PLEASE BE CAREFUL WHILE DRIVING OR WALKING IN THIS AREA. WATER MAIN IS BEING CLEANED AND LINED FOR

IMPROVED SERVICE

Pipe above ground is only temporary and will be removed as soon as possible.

Thank you, Your Water Utility

Figure 7-2   A sample caution notice to be posted at the work site

COMMUNITY RELATIONS

In addition to contacts with individual customers, overall community relations are equally important. Police and traffic control personnel and transit district representatives must be included in planning meetings to discuss detours and work schedules. The fire department also must be apprised of the project and any effects on emergency access to fire hydrants. Hydrants removed from service should be covered or otherwise marked to indicate their unavailability.

Emergency service providers such as paramedics or ambulance companies should also be alerted. Other public and investor-owned utilities, such as sewage, gas, electric service, and telecommunications (telephone, cable, and fiber-optics cable) should also be notified to allow coordination between the water line project and any other scheduled utility work. The local chamber of commerce or other local business groups may be contacted for assistance. Notices posted in store windows may help to avoid parking complaints.

SUMMARY

All notifications should include an emergency telephone number that is active 24 hours a day, 7 days a week. Good customer relations can be severely damaged by a customer's inability to report an emergency or discuss questions about water service. If a project requires entering private property to disconnect meters and hook up temporary lines, notices should clearly discuss the need for the work and how it will be conducted. As with any aspect of customer service, followup is extremely important and personnel must be available to respond promptly to all calls. Finally, an established policy and procedure for handling damage claims should be established.

AWWA MANUAL M28

Appendix A

Structural Lining Design Issues

STRUCTURAL CLASSIFICATION OF LINING TECHNIQUES

Lining systems used to rehabilitate potable water pipelines can be classified into four groups according to their effect on the performance of the lined pipe when subjected to internal pressure loads.

Class I Linings

Class I linings are essentially nonstructural systems used primarily to protect the inner surface of the host pipe from corrosion. They have no effect on the structural performance of the host pipe and have a minimal ability to bridge any existing discontinuities, such as corrosion holes or joint gaps. Hence, they have minimal effect on leakage. Their use is indicated in pipes suffering from internal corrosion or tuberculation but still in structurally sound condition and where abatement of current or future leakage is not an issue. Examples of Class I linings are cement–mortar lining and epoxy resin lining.

Class II and III Linings

Class II and III linings are both interactive and semistructural systems. When installed, the liners closely fit the host pipes, and any remaining annulus is rapidly eliminated when internal operating pressure expands the lining. Since the stiffness of such a lining is less than that of the host pipe, all internal pressure loads are transferred to the host pipe, leading to their classification as interactive. Such a lining is required only to independently sustain internal pressure loads at discontinuities in the host pipe, such as corrosion holes or joint gaps, or if the host pipe is subject to structural failure.

A liner is considered to be in Class II or III if its long-term (50-year) internal burst strength, when tested independently from the host pipe, is *less than* the

59

maximum allowable operating pressure (MAOP) of the pipeline to be rehabilitated. Such a liner would not be expected to survive a burst failure of the host pipe, so it cannot be considered as a replacement pipe. However, Class II and III liners are designed to bridge holes and gaps in the host pipe on a long-term basis, and various systems can be further classified in terms of the magnitude of the holes and gaps they can bridge at a given MAOP. Some systems are capable of bridging holes and gaps of up to 6 in. (52 mm) on a long-term basis at an MAOP of 150 psi (1,034 kPa).

The separation of these systems for spanning holes and gaps into two classes is based on their inherent resistance to external buckling forces and dependence on adhesion to the host pipe wall. Class II systems have minimal inherent ring stiffness and depend entirely on adhesion to the pipe wall to prevent collapse if the pipe is depressurized. Class III liners have sufficient inherent ring stiffness to be at least self-supporting when depressurized without dependence on adhesion to the pipe wall. As explained later, Class III liners can also be designed to resist specified external hydrostatic or vacuum loads.

Use of Class II or III linings may be indicated where the host pipe is suffering from one or more of the following conditions:

1. Severe internal corrosion leading to pinholes and leakage

2. Leakage from faulty joints

3. Localized external corrosion resulting in pinholes and leakage

Although the liner will not prevent further external corrosion, it will prevent leakage at corrosion holes. This capability guards against the associated effects of that leakage on the exterior of the pipe and the corrosivity and support offered by the surrounding soil.

Class IV Linings

Class IV linings, termed *fully structural* or *structurally independent*, possess the following characteristics:

1. A long-term (50-year) internal burst strength, when tested independently from the host pipe, equal to or greater than the MAOP of the pipe to be rehabilitated

2. The ability to survive any dynamic loading or other short-term effects associated with sudden failure of the host pipe due to internal pressure loads

Class IV linings are sometimes considered to be equivalent to replacement pipe, although such linings may not be designed to meet the same requirements for external buckling or longitudinal/bending strength as the original pipe. Also, they may be of smaller internal diameters. Class IV linings can, of course, be used in circumstances similar to those for Class II and III, but their use is essential for host pipes suffering from generalized external corrosion where the mode of failure has been, or is likely to be, catastrophic longitudinal cracking.

As explained later, some available renovation technologies can offer both Class II and III and Class IV linings, while a given lining system may be rated as Class IV for MAOP levels up to a threshold value and Class II and III for higher pressures.

ADDITIONAL DESIGN CONSIDERATIONS

In addition to internal pressure loads, linings may also be required to sustain external buckling loads during periods when the host pipe is depressurized, as well as transient vacuum loads. Some systems (Classes III and IV) can be designed to offer significant inherent resistance to such external loads, while others (Class II) depend solely on adhesion to the host pipe wall. Inherent resistance to external buckling normally varies with increased lining thickness and hence cost. Care should therefore be taken to ensure that such performance requirements are accurately defined.

The hydraulic performance of the lined pipe will be determined by the thickness of the liner, its closeness of fit to the host pipe, and its internal smoothness (C value). The lining process is usually preceded by extensive cleaning, which will itself restore the original flow cross section of the pipe. Liners of plastic materials are significantly smoother than the inner surface of a deteriorated host pipe, and they may even be smoother than the original pipe. In addition, many lining systems provide essentially joint-free coverage over long sections, so they offer less disturbance to flow than jointed sections of pipe. In general, close-fit plastic lining systems with SDR of 26 or more normally retain the original flow capacity of the pipe.

REINSTATEMENT OF FITTINGS AFTER STRUCTURAL LINING

All Class II, III, and IV linings block or otherwise interfere with the function of in-line fittings such as service connections, branch pipelines, and valves. Lining installers must remove branch connections, valves, reducers, and similar fittings at local excavations prior to lining. The liner then terminates on each side of the fitting location with suitable end seals, and crews then reinstate or install replacement fittings after lining.

Crews handle service connections in a similar manner. After accessing the connection fitting via a small local excavation prior to lining, workers disconnect the communication pipe, then they generally remove the fitting prior to lining. In an installation of a thick, polyethylene-based Class IV lining, workers commonly break out all or part of the existing pipe at the location of the service connection, line through the gap, and then install a new electrofusion or mechanical saddle connection.

For a Class II or III or thin Class IV lining, workers normally enlarge the hole left by the removal of the original connection, line over it, and then install a service connection adapter through the lining. Some lining technologies involve proprietary systems for such connections, and fittings manufacturers also offer ferrules suitable for a number of lining technologies. In some cases, ferrules can be installed under pressure so that new connections can subsequently be added to a lined pipe with minimum customer disruption.

End seals must be installed to prevent leakage between the liner and the pipe wall at any location where the liner is terminated. Both internal and external specialized seals are available. Some external seals incorporate flanges for easy reconnection to fittings. The seals vary in their capabilities for withstanding longitudinal forces and liner movement due to thermal and other effects, and these characteristics should be considered in selecting seals. Where sections of pipe have been removed to facilitate access for the lining process, a spool piece must be installed to bridge the gap after lining and end-seal installation.

The availability of reliable techniques for reinstating fittings is a vital element in the renovation of pipelines with structural linings. Options should be discussed in detail with contractors offering the lining systems. This choice also has implications for water companies, which must maintain inventories of specialized fittings for subsequent modifications or repairs to lined pipes. Procedures and costs required for repairs of lined pipes damaged by third-party actions should also be agreed with the contractor at the time of lining installation.

Index

NOTE: *f.* indicates figure; *t.* indicates table.

American Water Works Association
  cement–mortar lining standard, 25
  computer-based rehabilitation decision
    tools, 4
  disinfecting-pipelines standard, 21, 51, 53
American Water Works Association
 Research Foundation, 4
ANSI/NSF standards
  epoxy lining, 25, 26
  internal joint seals, 41

Bypass piping, 51, 54
  commercial installations, 52*f.*, 53
  problems, 53
  reconnecting service lines, 53
  residential installations, 51–53, 52*f.*

C factor, 3
Cable-attached cleaning devices
  drag cleaning, 10, 10*f.*
  electric scrapers, 11
  hydraulic-jet cleaning, 10
Cement–mortar lining, 21–22
  AWWA standard, 25
  machines, 22*f.*, 23*f.*
  operating procedures, 22–25
  return to service, 24–25, 25*f.*
CIPP. *See* Cured-in-place pipe lining
 techniques
Class I (nonstructural) linings, 4, 59
Class II/III (semistructural) linings, 4, 59–60
  and external loads, 61
  and hydraulic performance, 61
  and reinstatement of fittings, 61–62
Class IV (structural) linings, 4, 60
  and external loads, 61
  and hydraulic performance, 61
  and reinstatement of fittings, 61–62
Community relations, 58
Corrosion, 1–2, 2*f.*
Cured-in-place pipe lining techniques, 36
  classification, 36
  felt-based systems, 36–37, 37*f.*, 38*f.*
  membrane systems, 36, 39
  woven hose systems, 36, 39
Customer notification program, 55
  caution notices, 55, 57*f.*
  and detours, 55
  notification letter, 55, 56*f.*
Customer relations. *See* Community
 relations, Customer notification program

Disinfection, 21, 51, 53
Distribution systems
  hydraulic performance, 2–3
  structural performance, 3
  water quality issues, 1–2
Drag cleaning, 10, 10*f.*

Electric scrapers, 11
Encrustation, 1–2. *See also* Power boring
EPDM. *See* Ethylene propylene diene
 monomer
Epoxy lining, 25
  ANSI/NSF standard, 25, 26
  application head, 27, 27*f.*
  approved materials, 26
  machines, 26–27, 26*f.*, 27*f.*
  operating procedures, 28
  and precleaning, 25, 26*f.*
  return to service, 27, 27*f.*
Epoxy resin relining. *See* Epoxy lining
ERL. *See* Epoxy lining
Ethylene propylene diene monomer, 41

Fluid-propelled cleaning devices
  foam pigs, 11–14
  metal scrapers, 14–17
Flushing, 9
Foam pigs, 11, 11*f.*
  launching through fire hydrants, 13, 13*f.*
  loose debris flushed by water bypass, 11, 12*f.*
  operating procedures, 12–14
  pig launchers, 13
  progressive pigging, 11
  prover pigs, 13
  run sequence, 13–14
Fouling, 2

Hazen-Williams formula and coefficient, 2–3,
 3*t.*
HDPE. *See* High-density polyethylene
High-density polyethylene, 29–31
Hydraulic performance, 2–3, 3*t.*
  chart for selection of rehabilitation
    techniques, 7*f.*
Hydraulic-jet cleaning, 10

Internal joint seals, 41
  completion report, 45
  expanding into position, 44–45
  fitting procedure, 41–45
  joint filling, 42
  material details, 42*t.*

pipeline preparation, 42
retainer bands, 44, 44t., 45, 46f.
seal positioning, 43–44
seal testing, 45, 46f.
standard, 41
surface lubrication, 43, 43f.
surface preparation of joint area, 42, 43f.
valve (test unit), 43–44
widths, 41

Linings. *See also* Internal joint seals, Pipe
bursting
cement mortar, 21–25, 22f., 23f., 25f.
Class I (nonstructural), 4, 59
Class II/III (semistructural), 4, 59–60, 61–62
Class IV (structural), 4, 60, 61–62
cured-in-place, 36–39
epoxy, 25–28
modified slip-lining, 32–36
reasons for installing, 21
slip-lining, 28–32

*Maintaining Distribution-System Water
Quality*, 9
Maintaining service. *See* Bypass piping
MAOP. *See* Maximum allowable operating
pressure
Maximum allowable operating pressure,
59–60
Metal scrapers, 14–15, 14f., 15f.
operating procedures, 15–17
sandbag dams for particle settlement
ponds, 15, 16f.
in series, 15f.
spool pieces at entry and exit points, 15, 16f.
Modified slip-lining, 32. *See also* Slip-lining
classification, 32
distinctions from conventional slip-lining, 32
factory folded/hot re-rounded systems,
34–35, 35f.
folded and formed systems, 32, 33f., 34–36
and hydraulic cross section, 32
and liner thickness, 32
roller-based systems, 33
site cold-folded/cold re-rounded systems, 35
static die systems, 32–33
symmetrical reduction systems, 32–34

NFS International, 41

Pig launchers, 13
Pigs. *See* Foam pigs
Pipe bursting, 47
contractor certification, 49
and excavation, 48
hydraulic equipment, 47
patent licensing requirements, 49
and pipe diameters, 47

pneumatic equipment, 47, 48f.
and polyethylene replacement pipe, 49
and service connections, 48
static equipment, 47
testing and disinfection, 49
Pipes and pipelines
bypass, 51–54
cutting cast-iron, 24
disinfecting, 21, 51, 53
polyethylene, 49
preparation for internal joint seals, 42
Polyethylene pipe, 49
Power boring, 17
cleaning head, 17, 19f.
operating procedures, 19
rack-feed equipment, 17–18, 18f.
schematic, 18f.
Progressive pigging, 11
Prover pigs, 13
Public safety departments, 58

Rack-feed boring equipment, 17–18, 18f., 19f.
operating procedures, 19
Rehabilitation
renovation solutions, 4
replacement solutions, 4
selection of techniques for, 4, 5f., 6f., 7f.

Scrapers. *See* Metal scrapers
SDR. *See* Standard dimension ratio
Sedimentation, 1
Service continuity. *See* Bypass piping
Slip-lining, 28. *See also* Modified slip-lining
applications, 28
drawbacks, 31
operating procedures, 29–31
Standard dimension ratio, 29–30
Structural performance rehabilitation chart,
5f.
Stub ends, 31

Thermal butt fusion, 28
Thermoplastic pipe or liners. *See* Slip-lining
Tuberculation, 1. *See also* Power boring

Underwriters Laboratories Inc., 41
Utility companies, 58

Water mains
cleaning. *See* Cable-attached cleaning
devices, Fluid-propelled cleaning devices,
Flushing, Power boring
leakage and structural performance, 3
Water quality
chart for selection of rehabilitation
techniques, 6f.
issues in distribution systems, 1–2

AWWA Manuals

M1, *Principles of Water Rates, Fees, and Charges,* Fifth Edition, 2000, #30001PA

M3, *Safety Practices for Water Utilities,* Fifth Edition, 1990, #30003PA

M4, *Water Fluoridation Principles and Practices,* Fourth Edition, 1995, #30004PA

M5, *Water Utility Management Practices,* First Edition, 1980, #30005PA

M6, *Water Meters—Selection, Installation, Testing, and Maintenance,* Second Edition, 1999, #30006PA

M7, *Problem Organisms in Water: Identification and Treatment,* Second Edition, 1995, #30007PA

M9, *Concrete Pressure Pipe,* Second Edition, 1995, #30009PA

M11, *Steel Pipe—A Guide for Design and Installation,* Fourth Edition, 1989, #30011PA

M12, *Simplified Procedures for Water Examination,* Second Edition, 1997, #30012PA

M14, *Recommended Practice for Backflow Prevention and Cross-Connection Control,* Second Edition, 1990, #30014PA

M17, *Installation, Field Testing, and Maintenance of Fire Hydrants,* Third Edition, 1989, #30017PA

M19, *Emergency Planning for Water Utility Management,* Third Edition, 1994, #30019PA

M20, *Water Chlorination Principles and Practices,* First Edition, 1973, #30020PA

M21, *Groundwater,* Second Edition, 1989, #30021PA

M22, *Sizing Water Service Lines and Meters,* First Edition, 1975, #30022PA

M23, *PVC Pipe—Design and Installation,* First Edition, 1980, #30023PA

M24, *Dual Water Systems,* Second Edition, 1994, #30024PA

M25, *Flexible-Membrane Covers and Linings for Potable-Water Reservoirs,* Third Edition, 2000, #30025PA

M26, *Water Rates and Related Charges,* Second Edition, 1996, #30026PA

M27, *External Corrosion—Introduction to Chemistry and Control,* First Edition, 1987, #30027PA

M28, *Cleaning and Lining Water Mains,* First Edition, 1987, #30028PA

M29, *Water Utility Capital Financing,* Second Edition, 1998, #30029PA

M30, *Precoat Filtration,* Second Edition, 1995, #30030PA

M31, *Distribution System Requirements for Fire Protection,* Third Edition, 1998, #30031PA

M32, *Distribution Network Analysis for Water Utilities,* First Edition, 1989, #30032PA

M33, *Flowmeters in Water Supply,* First Edition, 1989, #30033PA

M34, *Water Rate Structures and Pricing,* Second Edition, 1999, #30034PA

M35, *Revenue Requirements,* First Edition, 1990, #30035PA

M36, *Water Audits and Leak Detection,* Second Edition, 1999, #30036PA

M37, *Operational Control of Coagulation and Filtration Processes,* First Edition, 1992, #30037PA

M38, *Electrodialysis and Electrodialysis Reversal,* First Edition, 1995, #30038PA

M41, *Ductile-Iron Pipe and Fittings,* First Edition, 1996, #30041PA

M42, *Steel Water-Storage Tanks,* First Edition, 1998, #30042PA

M44, *Distribution Valves: Selection, Installation, Field Testing, and Maintenance,* First Edition, 1996, #30044PA

M45, *Fiberglass Pipe Design,* First Edition, 1996, #30045PA

M46, *Reverse Osmosis and Nanofiltration,* First Edition, 1999, #30046PA

M47, *Construction Contract Administration,* First Edition, 1996, #30047PA

M48, *Waterborne Pathogens,* First Edition, 1999, #30048PA

M49, *Butterfly Valves: Torque, Head Loss, and Cavitation Analysis,* First Edition, 2001 #30049PA

M50, *Water Resources Planning,* First Edition, 2001 #30050PA

To order any of these manuals or other AWWA publications, call the Bookstore toll-free at 1-(800)-926-7337.

This page intentionally blank.